Standards, Quality Control, and Measurement Sciences in 3D Printing and Additive Manufacturing

Standards, Quality Control, and Measurement Sciences in 3D Printing and Additive Manufacturing

CHEE KAI CHUA
CHEE HOW WONG
WAI YEE YEONG
Singapore Centre for 3D Printing,
School of Mechanical and Aerospace Engineering,
Nanyang Technological University, Singapore

Academic Press is an imprint of Elsevier
125 London Wall, London EC2Y 5AS, United Kingdom
525 B Street, Suite 1800, San Diego, CA 92101-4495, United States
50 Hampshire Street, 5th Floor, Cambridge, MA 02139, United States
The Boulevard, Langford Lane, Kidlington, Oxford OX5 1GB, United Kingdom

Library of Congress Cataloging-in-Publication Data
A catalog record for this book is available from the Library of Congress

British Library Cataloguing-in-Publication Data
A catalogue record for this book is available from the British Library

ISBN: 978-0-12-813489-4

For information on all Academic Press publications visit our website at
https://www.elsevier.com/books-and-journals

Publisher: Matthew Deans
Acquisition Editor: Brian Guerin
Editorial Project Manager: Katie Chan
Production Project Manager: Julie-Ann Stansfield
Designer: Christian Bilbow

Typeset by Thomson Digital

CONTENTS

ABOUT THE AUTHORS

Chua Chee Kai is the executive director of the Singapore Centre for 3D Printing (SC3DP) and a full-time professor of the School of Mechanical and Aerospace Engineering at Nanyang Technological University (NTU), Singapore. Over the last 25 years, Prof. Chua has established a strong research group at NTU, pioneering and leading in computer-aided tissue engineering scaffold fabrication using various additive-manufacturing techniques. He is internationally recognized for his significant contributions in biomaterial analysis and rapid prototyping process modeling and control for tissue engineering. His work has since extended further into additive manufacturing of metals and ceramics for defense applications.

Prof. Chua has published extensively with over 300 international journals and conferences, attracting more than 6300 citations, and has a Hirsch index of 38 in the Web of Science. His book, *3D Printing and Additive Manufacturing: Principles and Applications*, now in its fifth edition, is widely used in American, European, and Asian universities and is acknowledged by international academics as one of the best textbooks in the field. He has also published two other books *Bioprinting: Principles and Applications* and *Lasers in 3D Printing and Manufacturing*. He is the world's number one author for the area of Additive Manufacturing and 3D Printing (or Rapid Prototyping, as it was previously known) in the Web of Science, and is the most "Highly Cited Scientist" in the world for that topic. He is the coeditor-in-chief of the international journal, *Virtual & Physical Prototyping* and serves as an editorial board member of three other international journals. In 2015, he started a new journal, the *International Journal of Bioprinting* and is the current chief editor. As a dedicated educator, who is passionate in training the next generation, Prof. Chua is widely consulted on additive manufacturing (since 1990) and has conducted more than 60 professional development courses for the industry and academia in Singapore and the world. In 2013, he was awarded the "Academic Career Award" for his contributions to Additive Manufacturing (or 3D Printing) at the 6th International Conference on Advanced Research in Virtual and Rapid Prototyping (VRAP 2013), 1–5 October, 2013, at Leiria, Portugal.

Dr. Chua can be contacted by email at mckchua@ntu.edu.sg.

Wong Chee How is an associate professor at the School of Mechanical and Aerospace Engineering, Nanyang Technological University (NTU), Singapore. He has extensive research experience of over 14 years in the area of modeling and simulation.

He received his Bachelor degree in Manufacturing Engineering from University of Birmingham, United Kingdom, and Master and PhD degrees from Nanyang Technological University, Singapore respectively. Prior to joining NTU in 2008, he was a senior research fellow in Data Storage Institute, A*STAR. In 2014, he was conferred fellow of the NTU Teaching Excellence Academy. Between 2014 and 2016, he served as the associate chair (Students) of the School of Mechanical and Aerospace Engineering. In 2016, he was as appointed associate dean (Academic) of the College of Engineering and is responsible for undergraduate and postgraduate curriculum, admissions, and teaching of the college. His research interests encompass simulation and modeling, such as atomistic simulations of materials at nanoscale and modeling of additive manufacturing processes, such as selective laser melting. To date, he has published more than 70 technical papers in international peer-reviewed journals, attracting more than 1000 citations. He currently has H-index of 18. He is now serving as the program director of the Future of Manufacturing program in the Singapore Centre for 3D Printing (SC3DP), SLM Solutions joint lab and NAMIC@NTU. He is an ExCo member of the American Society of Mechanical Engineers (Singapore) and served as its chairman from 2013 to 2014. He is also a member of ASTM F42 and Additive Manufacturing Technical Committee (Singapore).

Dr. Wong can be contacted by email at chwong@ntu.edu.sg.

Yeong Wai Yee is an assistant professor at the School of Mechanical and Aerospace Engineering, Nanyang Technological University (NTU), Singapore. She also serves as the program director for the Aerospace and Defence Programme in Singapore Centre for 3D Printing (SC3DP).

Assistant Prof. Yeong received her BEng (1st class Hons) and PhD degrees in Mechanical and Aerospace Engineering from the Nanyang Technological University in 2003 and 2006, respectively. Prior to joining MAE in 2013, she has industrial experiences in technical and supervisory functions in research and development, manufacturing, and quality systems. She has worked as a research engineer in SIMTech Singapore, associate research scientist at Abbott Vascular, research fellow at MSE NTU, and principal engineer at Alcon Singapore. Her job experiences allow her to have a comprehensive overview of the importance of standards and quality in manufacturing, as well as in research. Her main research interests are 3D printing, bioprinting, and the translational of the advanced technologies for industrial applications. Her current research topics include 3D printing of metal, multifunctional and lightweight structures, and bioprinting in tissue engineering. She has also published another book *Bioprinting: Principles and Applications*. She serves as an associate editor and reviewer in international journals. At current standing, she is within the world's top 25 most published scientists in 3D printing.

Dr. Yeong can be reached at wyyeong@ntu.edu.sg.

FOREWORD

Transformative technology is a rare thing, perhaps seen once in a generation. The acceptance of technology by the marketplace in question is largely dependent upon the degree of certainty associated with its application. How consistently it performs? How repeatable and reproducible are its results? How much confidence it instills within those who choose to use it? In many cases, the gap between a technology's research stage and its market application is vast, and in some cases never bridged. Wonderful and innovative advances may never achieve their full potential due to the lack of market access, in many cases determined by the quality, pace, and timing of a consensus standards program to build that elusive but critical bridge.

Additive manufacturing defined in ISO/ASTM 52900 as a process of joining materials to make parts from 3D model data, usually layer upon layer, as opposed to subtractive manufacturing and formative manufacturing methodologies, ranks high on the transformative scale.

In recent years, use of additive technologies has evolved within a variety of mainstream manufacturing sectors, including aerospace, automotive, biotechnical/medical, and consumer products. Despite this growth, the pace at which consensus standards (and the measurement science upon which they are based) are being developed hasn't mirrored the accelerated evolution of the technologies and their ever-expanding industry applications. In response to this disparity, accelerated standards development models are becoming increasingly necessary to satisfy needs associated with the expansion of additive manufacturing methodologies into these new market segments.

There is so much work to be done. Given the relatively finite collection of technical experts with a sufficient depth of knowledge and experience to identify and develop these much-needed standards, collaboration and its resulting lack of duplication becomes critical. Leading the pack in the standards space is ASTM International Committee F42 on Additive Manufacturing Technologies and ISO TC261 on Additive Manufacturing. Via a Partner Standards Development Organization (PSDO) agreement between their parent organizations, these global standards developers have maximized their bandwidth in both process and expertise applications. The marriage of horizontal and vertical relevance inherent in the infrastructure of the respective committee populations has created an environment ideally suited to tackle the challenges associated with "standardizing" an area in dire need of a high quality and market relevant bridge. Needs currently being addressed to include the ability to comparatively benchmark the performance of different additive processes, improvement of the buyer/seller dynamic by enabling increasingly accurate specification of production requirements, and offering the research community a mechanism to provide repeatable results capable of independent verification.

Standards come in many shapes and sizes, and one size rarely fits all. Within the world of additive manufacturing, the types of standards needed are almost as diverse as the applications served by the AM technologies. Examples of standards include specifications, test methods, classifications, practices, guides and terminology. Specifications are needed for physical, mechanical, or chemical properties, as well as safety, quality, or performance criteria. Test methods are critical for identification, measurement, and evaluation of one or more qualities, characteristics, or properties. There is a need to create Classifications of materials, products, systems, or services into groups based on similar characteristics, such as origin, composition, properties, or use. Practices should be standardized for application, assessment, decontamination, inspection, installation, preparation, sampling, screening, and training. Guides will be helpful to increase the awareness of information and approaches in a given subject area. Standardized terminology is necessary to establish a uniform and consistently applied language for industry consumption.

The value of a consensus standards program is beyond contestation, but within the rapidly evolving world of additive manufacturing, not just any program will do. To be truly effective, standards development must be nimble, flexible, open, transparent, living, and timely. The marriage of these attributes creates an environment where standards development for additive manufacturing technologies will thrive, putting to rest the contention that standards stifle innovation. Quite to the contrary, a robust, inclusive, and well fleshed out standards program will enable innovation by creating an environment where measurable points of distinction don't come from simply meeting the standard, but via exceeding it and by how much. Competition such as this drives innovation. Innovation expands markets. Markets thrive under well-constituted standards programs.

Standards are multifaceted things. While based on sound and reliable science possessing enormous technical sophistication, they are also, in many respects, enablers and gatekeepers of market access. This combination of technical and economic relevance is incredibly powerful, and is vital to the transformative technology, that is, additive manufacturing achieving its maximum potential.

Pat A. Picariello, J.D., CStd
Director, Developmental Operations, ASTM International

PREFACE

Additive manufacturing, also known as 3D printing, is revolutionizing the landscape of manufacturing, creating new opportunities in design and generating new routes of fabrication for various products. In recent years, additive manufacturing plays a strategic role in the future of manufacturing for both the consumer products and high value products. 3D printers are installed in households, classrooms, and in the production floor of high-tech manufacturing sights. This new technological trend is even more evident in highly regulated industries, such as biomedical and aerospace industries. Quality systems and quality assurance are prerequisite for successful adoption of 3D printing into the value chain of these services and products.

On the research front, additive manufacturing technology is creating new paradigm in research fields, such as bioprinting, electronic printing, as well as environmental-related fields. These are highly translational technology domains, which require a concerted effort between different institutes and collaborations across different disciplines. Therefore, it is equally critical to apply the principles of quality management coupled with reliable measurement techniques when creating new sciences in additive manufacturing research.

However, there has been limited progress in the standard and measurement sciences in additive manufacturing. This textbook aims to present the critical components and current progress of standards and measurement sciences in additive manufacturing. This will be the first textbook on standards, quality, and measurement sciences in additive manufacturing.

Through this textbook we aim to achieve fundamental understanding in these aspects:

- The importance of standards and measurement sciences in additive manufacturing.
- The current landscape of standards in additive manufacturing.
- The quality framework tailored for additive manufacturing processes.
- The data format and process control in additive manufacturing.
- Overview of different materials and characterization methods.
- Overview of equipment qualification activities and safety consideration regarding to different 3D printing systems.
- Benchmarking and metrology for 3D printing systems and printed parts.

This book is suitable for students, researchers, engineers, general public, and regulatory specialists from all industry sectors.

Chee Kai Chua
Chee How Wong
Wai Yee Yeong

ACKNOWLEDGMENTS

First, we would like to thank our respective spouses, Wendy, Yin Chee, and Tee Seng, and our respective children, Cherie Chua and son-in-law Darren Lee, Clement Chua and daughter-in-law Lynette Balakrishnan, Cavell Chua, Lim Bao Rong, and Lim Zi Kai for their patience, for their support and encouragement to complete this book.

We are grateful to the administration of Nanyang Technological University (NTU) for valuable support, especially from the Singapore Centre for 3D Printing (SC3DP), and the School of Mechanical and Aerospace Engineering (MAE).

In addition, would like to thank the following colleagues, for their valuable contributions. They are: Dr. Lee Jian Yuan, Dr. Loh Loong Ee, Dr. Sing Swee Leong, Goh Guo Dong, Goh Guo Liang, Gregory Chua Kok Hong, Joel Tan Heang Kuan, Joel Lim Choon Wee, Lu Qingyang, Peh Zhen Kai, Rahul Koneru, Tey Cher Fu, and Yap Yee Ling.

We wish to express sincere appreciation to our special assistant Kum Yi Xuan for selfless help and immense effort in the coordination and timely publication of this book. We would also like to extend our special appreciation to Pat A. Picariello, Director of Developmental Operations at ASTM International for his foreword.

Chee Kai Chua
Chee How Wong
Wai Yee Yeong

CHAPTER ONE

Introduction to 3D Printing or Additive Manufacturing

Contents

Standards, Quality Control, and Measurement Sciences in 3D Printing and Additive Manufacturing
http://dx.doi.org/10.1016/B978-0-12-813489-4.00001-5

1 INTRODUCTION TO ADDITIVE MANUFACTURING

In recent years, the upgrading and replacement of products have become increasingly rapid. New products with enhanced functions and/or more innovative designs supersede existing products in the market. The increased competition for manufactured products to reach the global market before any competitors has resulted in companies having to launch their new products in the shortest possible time. Conventional manufacturing technologies, in general, involve long production times, are inherent to material wastage due to the subtractive nature of the processes and are craftsmanship intensive. Some examples include casting and machining. In order to meet the demand in this accelerated process of product change, new technologies have to be developed. Time spent on the design, manufacture, test, and market phases have to be shortened.

Additive manufacturing (AM) or 3D printing, as its name suggests, fabricates parts by adding successive layers of material under computer control [1–3]. A computer-aided design (CAD) is created and exported to stereolithography (STL) file format that is read by the AM equipment. There are many techniques available, which can be categorized according to their raw material. They are: (1) powder-based, (2) liquid-based, and (3) solid-based. Some examples of powder-based techniques include selective laser melting (SLM), selective laser sintering (SLS), and electron beam melting (EBM). Liquid-based techniques include stereolithography apparatus (SLA) and polyjet while solid-based techniques include laminated object manufacturing (LOM) and fused deposition modeling (FDM). The strengths of AM compared to conventional manufacturing methods are listed further.

1.1 Small-Scale Production

AM is primarily targeted at small-scale customization and personalization while conventional manufacturing is skewed toward mass production. 3D printing in small quantity is also cheaper as compared to conventional techniques, where tooling and preparation procedures prior to actual production are intensive and costly. One of the greatest business opportunities for customization is in the healthcare industry. Implants currently come in several fixed sizes and should the patient fall in between any two, he/she will feel uncomfortable due to the uneven fit. Another example is the case with dental crowning. The dentist takes an impression of the tooth for crown making. The product, however, is usually of imperfect fit and manual crafting is needed to improve the fitting of the crown. Hence, experience and excellent craftsmanship are required for a successful procedure. Now with AM, customization is possible and relatively effortless.

1.2 Lower-Cost Production

Unlike AM, conventional processes are usually labor intensive in the operational, assembly, and inspection stages. Great expertise is required to operate traditional processes successfully as they are error-prone and require post-processing. Most of the material removed from bulk is wasted, as it cannot be directly reused like in some AM processes. For example, up to 90% of the bulk material is removed in the creation of an aircraft component through conventional manufacturing. With AM, however, raw material utilization is increased and wastage is greatly reduced. Labor, material, and tooling costs incurred in traditional processes are significantly higher. Assembly labor can now be eliminated when sub-assemblies can be 3D printed into a single part. Moreover, there is also an advantage in lead-time for raw materials used in AM. For example, titanium powder used in AM only needs to be ordered 3 weeks in advance, while the same material in billet form used in conventional manufacturing will have to be purchased a year or two prior to its usage. This is because billets are cast to required size and microstructure, and have to be produced in large quantities for economy of scale, as compared to powder which is premanufactured given its generic form and is readily available. Therefore, this lowers inventory costs and overall production time, influencing the adoption of AM technologies in businesses.

1.3 Responsive Production

The production process is highly reactive and versatile. Designs can be printed, tested, modified, and reprinted quickly. One example of such use is in Formula One racing. Engineers are printing racecar components and analyzing the performance of the newly printed parts, while an improved version is already being prepared for printing. Conventional techniques often require a long time to manufacture a particular component as compared to AM. For example, the production of an aircraft component through the former techniques takes up to 60 weeks but only a month with the latter.

1.4 Shorter Supply Chains

Currently, in the conventional manufacturing industry, products are distributed and stored in warehouses before being delivered to customers as and when needed. Additional costs incurred by an unsold product increases over time. Advanced logistics are also critical in identifying the cheapest route to a particular product. If the shipment of an outsourced product can be fulfilled within a reasonable lead-time and the cost to do so is cheaper than having it fabricated locally, the product will still be shipped even if it is half way around the globe. Today, made-to-stock strategy is still more prevalent than made-to-order manufacturing process. AM is capable of shortening the supply chain as companies can now directly fabricate any components on site. Shipping costs and carbon footprint are reduced through the elimination of transportation, making manufacturing closer to the customers than ever before. As the cost of a 3D printer decreases while labor and oil prices increase, the adoption of 3D printing technologies in manufacturing can, therefore, break the traditional concept of manufacturing.

1.5 Optimized Design

AM has the capacity to fabricate complex designs capable of enhancing the performances of their applications. One such example is the fabrication of lattice structures used in the automotive industry intended to enhance the mechanical impact response with an overall reduction of weight. The ability of AM in allowing reiteration and rethinking of the fundamentals and reconstruction of products bring about the creation of new and improved merchandise. On the contrary, there are limitations to the level of component complexity when one wants to fabricate within a reasonable budget and quality through conventional manufacturing. Some examples include the need for draft angles to ensure easy ejection of the fabricated part; corners should be designed with radiuses in order to reduce stress concentration; and volume shrinkage has to be taken into account prior actual manufacturing so as to achieve dimensional accuracy.

2 EXPANDING ROLE OF ADDITIVE MANUFACTURING

Strategic foresight is important when the global environment is getting increasingly complex and dynamic. As such, it is necessary to think ahead and predict any possible changes. Some key factors in the future of manufacturing with AM are in the customization of part design, integration/automation of the production processes with AM technologies, AM of tailor-made part properties, acceptance of AM standards and certification of AM technologies. There is an increasing usage of AM technologies in the production of aerospace, automotive, and electronic components. Certification institutions have recognized the importance of AM technologies and the need to develop certification processes for newly fabricated AM products, fostering higher transparency and increasing trust in AM techniques as a safe and reliable technology. Although the market penetration by AM is still currently limited, there are many possible areas to expand the application of AM technologies. One example is to integrate AM technologies with electronics production processes to achieve better advancement in the electronics industry.

3 MATERIALS IN ADDITIVE MANUFACTURING

One of the early applications of AM has been to rapidly produce net-shaped plastic prototypes as no expensive special tools or skill intensive tooling works are required. Its ability to produce complex structures of near net shapes in materials that are conventionally hard to machine has gained this manufacturing technique much popularity. Such materials include hard metals, ceramics and composites. The improvement and development of materials used in AM have been subjected to tremendous rapid progress. Toxic acrylic photo-resins have been substituted with better performing epoxy-based ones in SLA. Instead of just nylon and wax, acrylonitrile butadiene styrene (ABS) can now be used in FDM. With SLS, it is now possible to sinter metals or ceramics without the use of polymer binders.

Basically, any material can be produced by one or another AM technique today. These materials can be divided into four main categories: plastics, metals, ceramics, and composites.

3.1 Plastics

A broad range and successively increasing variety of plastics suitable for AM are available in the market. These materials vary in transparencies, thermal or mechanical properties. Depending on the application of the printed plastic product, a suitable material should be selected. Currently, materials exhibit properties mimicking those used in traditional manufacturing and have since replaced poor performing brittle materials used in the early 1990s.

Polyamides are the most popular thermoplastics used in plastic laser sintering due to their widespread use in injection molding. However, the specific grades of polyamides used in AM have different physical properties and wider processing windows as compared to their injection molding counterparts even though they are chemically identical. Products fabricated through laser sintering also have significantly different material properties from those that are injection molded. Not only is this due to the difference in material properties, but is also caused by the differences in operating conditions, such as pressure and cooling rate experienced by the polymer during the two different fabrication processes.

Mechanical properties of polyamide-based powders used in plastic laser sintering can be improved. One such example is the use of glass-filled powders. This is analogous to fiber reinforced injection molded materials, where such reinforcements improve the stiffness, temperature resistance, and isotropic properties of the product. As the number of AM systems installed increases worldwide, third parties specializing in powder production influence the economics and drive the development of new materials.

Any material can be used in plastic laser sintering as long as they are polyamide-coated and are spherical for good powder flowability. The coating acts as a binder with the internal spheres acting like granules. Polystyrene is also used in plastic laser sintering, preferably for applications where toughness, impact resistance, and molded-part performances are required.

FDM uses thermoplastics such as ABS, polycarbonate (PC), nylon, and polyphenylsulfone that are similar to those tried and tested in traditional manufacturing processes. These materials come in varying tolerances, mechanical, and chemical properties and environmental stability with specialized properties such as electrostatic dissipation, translucence, biocompatibility, or flame tolerance.

Liquid resins that mimic the properties and aesthetics of polypropylene and ABS are available in SLA. New composite photopolymers with nanosized particles are developed to improve the overall performance of the product. Apart from the varying mechanical properties, such as stiffness and heat deflection in liquid resins, volume shrinkages, and the ease of processing are also improved.

Today, a variety of plastics with vastly different mechanical, chemical and environmental properties can be additively manufactured, except for imidized materials. Polyimides are a group of polymers with exceptionally high heat and chemical resistance that are yet to be used in AM. It will be interesting to see the utilization of such materials in AM.

3.2 Metals

The most frequently used techniques in AM of metallic parts are divided into partial-melting processes and complete-melting processes. The former includes SLS [4,5] and laser microsintering while the latter comprises SLM [6–9], 3D laser cladding and EBM. The densities of parts fabricated by partial-melting techniques vary between 45% and 85% of the theoretical density. Furnace sintering and infiltration of the component with a material of a lower boiling point is usually required to increase the final density, making it a two-step process. On the other hand, complete-melting techniques are capable of building parts of high density, comparable to traditionally manufactured ones. Metallic powders used are usually spherical with an average diameter of 45 microns. Depending on the type of metal powder used, the laser beam diameter of the system, layer thickness selected, and the desired integrity of the fabricated part, a suitable set of process parameters that include laser power, scan speed, and scanning strategy should be identified. Heat treatment may be applied to reduce the thermal residual stresses or optimize the microstructure of the fabricated metallic parts. Moreover, further post processing may be necessary to achieve high geometrical accuracy and high-quality surface finish in these processes.

The most commonly used metals in AM are steel and its alloys due to their availability, reasonable cost, and biocompatibility as bone and dental implants. Titanium and its alloys are less commonly used followed by nickel, aluminum, copper, magnesium, cobalt-chrome, and tungsten. Precious metals such as gold are also used as summarized in Table 1.1. Metallic parts produced by AM generally exhibit better mechanical properties than conventionally manufactured counterparts due to its higher cooling rate, which results in finer microstructure and grain size that enhances the mechanical properties.

3.3 Ceramics

Ceramics are widely used in AM as shown in Table 1.2. They were used as a powder-binder combination in SLS and as a ceramic filled resin in SLA. Furthermore, they were also explored to be used in SLM to produce high-density ceramic parts. However, AM of ceramics is more challenging than metals due to their high melting point and low formability. They are also susceptible to thermal shocks that result in cracks in the printed ceramic parts which can be mitigated by preheating the powder bed. Furthermore, ceramic powders have poor flowability that inhibits smooth depositions of powder layerswhich can be improved with spray drying. CO_2 lasers are often used to melt ceramics as they exhibit higher absorptivity under this wavelength.

3.4 Composites

Crack-free metal matrix composites (MMC) of 99.9% density can be coupled with tungsten carbide-cobalt (WC-Co), ceramic or nonferrous reinforcements to enhance the mechanical properties. Such 3D printed composites are usually used in extreme environmental conditions, which include the oil and gas, mining, automotive, or power industry due to its high

Table 1.1 Types of metal alloys used in AM

	Type of alloy	References
Iron	316L stainless steel	[10–22]
	314S stainless steel	[12,23–25]
	304L stainless steel	[26]
	Inox904L stainless steel	[18,27,28]
	M2 highspeed steel	[11,12,23,29]
	H13 tool steel	[23,30–32]
	H20 tool steel	[33]
	Maraging steel	[19,34–39]
	Precipitation hardening (PH) steel	[40–42]
	Austenitic and martensitic steel mixture	[43]
	Fe-Ni alloy	[44,45]
	Fe-Al alloy	[46,47]
	Fe-Cr-Al alloy	[48]
	Fe-Ni-Cr alloy	[45,49]
Titanium	Commercially pure titanium (cpTi)	[50–58]
	Ti-6Al-4V (Ti64)	[42,54,59–62]
	Ti-6Al-7Nb	[63–65]
	Ti-24Nb-4Zr-8Sn	[61,66]
	Ti-13Zr-Nb	[67]
	Ti-13Nb-13Zr	[68]
Nickel	Pure nickel	[69]
	Inconel 625	[30,31,70–72]
	Inconel 718	[73,74]
	Hastelloy X	[75–77]
	Nimonic 263	[78]
Aluminum	Pure aluminum	[79]
	AlSi10Mg	[69,80–86]
	Al6061	[17,81,87]
	AlSi12	[81,88,89]
Copper	Pure copper	[90]
	C18400	[91,92]
	CuNi15	[45]
	Other compositions	[13,17,20,31,92,93]
Magnesium	Pure magnesium	[94–96]
	Mg-Al	[79,97]
Cobalt-chrome (Co-Cr)		[98,99]
Tungsten		[94,100–102]
Gold		[103,104]

Table 1.2 Types of ceramics used in AM

Types of ceramic	References
Alumina	[105]
Alumina-zirconia mixture	[106–108]
Alumina-silica mixture	[109]
Silica	[110–112]
Silicon carbide	[113]
Silicon monoxide	[113]
Yttria-stabilised zirconia (YSZ)	[106,114]
Tricalcium-phosphate (TCP)	[114]
Li_2O-$Al2O_3$-SiO_2 (LAS) glass	[115,116]

hardness and wear resistance. The uniform fine microstructure contributes to the increased hardness, eliminating any need for further improvements in mechanical properties through costly postprocessing or heat treatment procedures. MMC powders are usually prepared by mixing different powders of interest. When the metal particles melt during the printing process, the molten metal matrix will bind the structural reinforcements together. AM of MMC has gained much attention due to its ability to fabricate composites with unique properties that cannot be found in mainstream metals or alloys. SLS has the ability to yield structures comparable to conventionally manufactured hard metals and the most extensively studied composite in AM is the WC-Co MMC. Laser metal deposition (LMD) has also been used to create bulk WC-Co MMC from ball-milled powders that comprise WC crystallites in Co matrix. Variation in microstructure with specimen height can be observed in LMD printed parts. This is due to the inherent difference in cooling rates experienced by the material during the fabrication process, which causes a change in the hardness with specimen height.

Some challenges faced in the 3D printing of fully dense MMCs of desired homogeneous microstructures are the entrapment of gases, the presence of microcracks at the metal matrix and reinforcement interfaces and particulate aggregation. Debonding at the interfaces is detrimental to the mechanical properties of MMCs as they become sources of crack propagation. One example is the bending test of SLS fabricated nonferrous MMC that consists of titanium-carbon/iron, nickel (TiC/(Fe, Ni)) reinforcements. The composite is ductile on the overall but brittle at the metal matrix and ceramic TiC interface as metals and ceramics have poor wettability. An effective way to overcome this issue is to modify the interfacial structure by coating ceramic particles with a layer of metal to enhance wettability. Ceramic TiC particles coated with nickel can be used to reinforce

Inconel 625 and Ti-6Al-4V metals, preventing the formation of micro-cracks between metal-ceramic interfaces and aggregations of ceramic particles that causes nonhomogeneity. Additions of rare earth (RE) compounds have also been reported to improve the processability of MMCs due to its ability in enhancing particulate dispersion that gives rise to the fabrication of homogeneous fine microstructures.

4 APPLICATIONS IN ADDITIVE MANUFACTURING

It is often misunderstood that a particular AM technique is exclusive to fabricating components for a specific type of application. This is not true. There are many different AM techniques that are capable of solving similar problems. Apart from the intended applications, special requirements are taken into consideration when selecting a suitable material. The design dimensions, dimensional accuracy, surface roughness, resolution quality, and temperature range that will be experienced by the fabricated component are some factors that can affect the selection of material. A suitable AM technique capable of processing the selected material will then be chosen.

AM can be used for direct or indirect prototyping, manufacturing and tooling purposes. Direct processes use AM techniques to print digital models into physical components directly. However, as there are restrictions to the material color, transparency, and flexibility that can be used in material dependent AM, not all designs can be 3D printed. Not only so, compared to conventional manufacturing techniques, AM is primarily used for small-scale production. It is expensive to print in large quantities. Therefore, a master copy of the design is first printed before being used for mass production through cheaper alternatives. Such processes are termed indirect as the products are mass-produced through nonadditive techniques. Direct or indirect processes aside, AM finds its applications in automotive, aerospace, medical, educational, and casting industries.

4.1 Automotive Industry

Since the late 1980s, AM has been adopted by the automotive industry for designing and redesigning of car components. A diverse range of prototypes has been printed at the product development stage and are used for negotiation and evaluation purposes. As customization becomes increasingly important, AM is an attractive solution due to its abilities to tailor made products that conventional techniques find difficulties in delivering.

AM was once used for printing prototype of car interiors intended for testing and communication purposes, while conventional techniques, such as plastic injection molding were used to manufacture actual car components. The interior design of a car often influences its sales and now with customizable interiors, buyers can personalize according to their taste and budget, boosting car sales. As AM is capable of meeting the demands of variations, printed parts now find their way to actual use in a car more frequently. Dashboards of exact dimensions are printed in smaller sections before being glued together and coated with acrylic paint. The high printing accuracy of SLA allows the seamless fitting of separate sections and the fabrication of functional fitted parts. An actual working airbox of an engine can also be printed through SLA. Due to the high temperature experienced during combustion, thermoset polymers capable of withstanding the tremendous heat are selected for printing.

It is not always economical to produce parts conventionally especially if the production volume is not large enough. Short series injection molding inserts are more cost efficient to be printed through SLA instead. This technique prints mold inserts with high dimensional accuracy and by using ABS-like materials with high toughness, the printed mold can withstand the forces and heat experienced during injection molding. Some examples of car exterior parts that can be printed and used directly out of the 3D printer are the front fenders of semi-trucks and customized duct outlets. These components can be fabricated within a short timeframe through additive techniques such as SLA and FDM where a large part can be printed in smaller sections before being assembled together, coated with the desired color and decorated accordingly.

National Aeronautics and Space Administration (NASA) engineers are 3D printing up to 70 parts of their space rovers for explorations on asteroids and eventually, on planet Mars. AM provides a simple and effortless alternative in the fabrication of certain space rover components for what would otherwise be a complex process to conventionally manufacture. More often than not, designs are not exact matches to what the designers intended. However, with AM, it is now possible to print designs in a day or even several hours. The actual physical design can then be checked and modified quickly. Given the tight project budget and period, AM makes the designing process much more cost effective and time efficient as compared to conventional manufacturing techniques.

4.2 Aerospace Industry

Currently, AM is widely used in all stages of the aircraft manufacturing industry, from the initial design to the testing, tooling phases, and finally to the actual production of an aerospace component. Not only so, AM is also used for repair works and as support systems. The increasing adoption of AM techniques can be observed among aerospace original equipment manufacturers (OEMs), maintenance, repair, and overhaul (MRO) companies. Special design requirements of a component or complete assemblies are often desired in aerospace industries. These needs can easily be met by the agility of AM and its empowerment in small and medium-sized enterprises allow them to compete with the big industries.

Aerospace industries are leveraging on the advantages in change and innovation expeditions AM has to offer. Another reason for its growing attractiveness in the aerospace industry is its ability to print lightweight structures using flame and chemical retardant materials. Aerospace components are required to withstand harsh conditions such as extreme temperatures and G-force stress fractures. With the development of highly resistant materials, AM sees an increase in its applications. Some examples are the fabrication of tooling, fixtures, and actual end products used in the aircraft. AM allows the printing of tooling needs for aerospace applications using composite materials. The cost and time of production are much lower than traditional manufacturing, which requires several thousands of dollars and months to complete. The versatility of AM allows the luxury of design modifications with no significant rise in cost and time delays.

Composite materials can be wrapped around 3D printed soluble cores to create hollow composite components. One example is the fabrication of an unmanned vehicle capsule. AM eliminates the need for tooling works in repair and small-scale manufacturing which often makes up the majority of the overall cost. Compared to conventional manufacturing methods, manufacturing tools such as molds, jigs, surrogates, fixtures, and templates can be printed for use in a matter of hours, not weeks. One example is the building of polycarbonate wiring conduits through FDM in less than three days at a small price. A similar aluminum cast component, however, will take more than 6 weeks to fabricate. Furthermore, airlines have different requests for their aircraft interior. As such, Boeing, an aircraft manufacturing company, is printing the customizable interiors of its aircraft. It is time-consuming to fabricate such customizations conventionally and the high cost to produce a few copies is unjustifiable. General Electric Aviation is also using 3D printing technologies to reduce engine weight, saving on fuel consumption.

4.3 Medical Industry

Biocompatible materials of varying material properties can be used to print surgical instruments and prototypes with great dimensional accuracy. An example of such application is the fabrication of customized fixtures or strong tooling using high-performance thermoplastic materials through FDM. Since 2008, 3D printing has been used by doctors at the aesthetic plastic surgery center in Singapore's National University Hospital (NUH). 3D models of the patients' skulls were printed as aids for surgery, providing doctors with precise visual information that could not be attained otherwise. Such important information better prepares the surgeon prior to the actual surgery. To replicate part of a damaged skull, a 3D profile of the area is scanned and converted into data. Printed skull implants fit seamlessly with the patients due to the high dimensional accuracy and smooth surface finishes.

With design flexibility, customizability and high product quality, AM is also adopted in building parts of the robotic exoskeleton for patients born with arthrogryposis multiplex congenita (AMC). Patients born with AMC have stiff joints and underdeveloped muscles. These patients do not gain the ability to lift their arms by their own strength even as they grow. Such inabilities in using their limbs can affect development, leading to limited cognitive and emotional growth. Traditionally, robotic arms are made of metallic parts that are too heavy and bulky for children. However with 3D printing, durable, light, and small components appropriate for daily usage fit for a child can be customized and printed. Should the robotic arm break during the course of its use, another replica can easily be printed with no significant lead time. This empowers young patients with the ability to carry out daily activities, such as playing and feeding themselves.

AM is also used in orthodontic treatments that are traditionally laborious. The number of alterations required to successfully complete a particular procedure is highly dependent on the experience of the dentist and communication with the patient. Instead of uncomfortable oral impressions taken in traditional practices, digital images of the oral cavities are scanned for 3D printing. Oral implants such as retainers, aligners, and expanders can be printed quickly, easily, cleanly, and painlessly with high accuracy and precision. Since all of the information is stored digitally, it is no longer necessary to store bulky physical models for years. This decreases the labor cost, the workload of and possible human errors incurred by the dentist. Accuracy, efficiency, and success of a treatment procedure are dramatically increased by shortening production time and increasing output. As such, both patients and dentists will have more free time when appointment durations and frequencies are shortened. Investment in digital orthodontics was once too expensive and inaccessible for smaller laboratories and clinics. Now, there are professional easy-to-use 3D printers that are compact and office friendly that are available at an affordable price. This increases the productivity and revenue of dental clinics through patient growth while streamlining the implant production process, offering a cleaner solution for orthodontic laboratories.

4.4 Education

AM enables students to realize their ideas through the fabrication of the actual physical product, stimulating, and driving their creativity. The exposure of this futuristic technology to young learners in school better prepares them for the challenges of the future. A larger driving force is often required to excite students above the age of 11 to learn about design and engineering. The 3D printing process allows students to see the formation of their designs and this often amazes them. An example of the implementation of AM is at STARBASE Minnesota, a program with curriculum aligned with national standards. Students have a fresh take on mathematics, science, design, and engineering learning in an integrated fashion through specific missions, such as the launching of a rocket model. Students are having so much fun in solving their work without even realizing that they are doing mathematics and science. Students are tasked to design rocket fins on CAD systems, 3D print them and launch the rocket to collect the performance data before making discussions on how features of their rocket fins affected the performance. This provides students with a realistic learning opportunity, as the process is similar to what real engineers do at work.

Educators themselves are attending workshops where they are exposed to different kinds of 3D printing technologies, and learn about the latest methods, materials, and happenings in the industry. Their hands-on experience on the technology is important in the smooth impartation of their skills to the students, better preparing them for real work problem-solving skills. 3D printing is not only used for educational purposes but also for high-end research especially in the field of engineering and architecture in universities. Students are using 3D printers to fabricate new forms of prototypes that are too complex to be produced otherwise. Designs requiring small features that other technologies cannot produce are now achievable with AM. One example is the redesign of a traditional aluminum car bumper that requires a day or two to be manufactured conventionally. Making use of the advantage of 3D printing in fabricating complex structures, students created a plastic design that is just as lightweight and strong as the conventional design for their robotics application, before printing it out in less than two hours. Moreover, the high resolution of 3D printers allows the fabrication of fine details in their design.

Research students are also using AM to produce anatomically true functional human elbow for robotics research. The material properties of the complex design can be tuned and printed in multimaterial, something that cannot be achieved with conventional manufacturing. Having valuable experience in 3D printing better prepares and gives students a competitive advantage in their job search. Manufacturing constraints that have been present for decades no longer apply. Students have to rethink the way they design products and 3D printing is the technology that allows the realization of complex novel designs. Young students are problem solvers of the future and exposing them to AM equips them with the tools necessary in building the bright future.

4.5 Prototyping

Manufacturers equipped with 3D printers find themselves having the capabilities to grab demanding business opportunities that cannot be achieved with conventional manufacturing techniques. The flexibility of AM allows manufacturers to amend the product design at any point of time by simply editing the CAD file. One example is the revolution of conventional injection molding techniques with AM, focused on small to medium sized prototyping business opportunities. Tooling costs of injection moldings are high; the development and building process is time-consuming. With AM, it is possible to take a CAD file, import it to the machine and print a part with in a matter of hours. Even if it is a large part with complex geometries, it will only take up to a maximum of several days to complete, not weeks and months like in traditional manufacturing. Competition in the traditional manufacturing sector is stiff but with 3D printing, companies no longer have to secure businesses through quick production turnaround but rather in the development, designing, proofing, and prototyping phases of the idea. This allows the companies to help clients realise an idea through prototyping development and production.

More often than not, companies have to produce results within tight timelines. Products that require modifications can be studied, improved and printed immediately, all within a short period of time. Any errors in the printed prototype, such as mismatches in the assembly of the components can be corrected and reprinted right away. If such errors are not corrected before sending it for mass production, the product will not be tool safe and would not have worked. This would cause the tool to be scraped, rebuilt and incur extra time and cost. 3D printing technologies are therefore a game changer, bringing about a difference between killing a program to keeping it moving forward. Companies that have invested in AM technologies to make their processes better are no longer competing with overheads, mark-ups, debts, and lead times. This provides a competitive edge by rendering different services, which differentiates themselves from their competitors.

FDM is an example of a frequently used AM technique in the prototyping of medical equipment, such as injection systems that deliver contrasts through the injection of dyes into the vascular system. This allows physicians to visualize the vasculature anatomies and to eventually develop therapeutic solutions. Companies are constantly trying to increase their efficiencies in bringing products to their customers before the competitors. With 3D printing, the creative process of product development is facilitated, and hence the designs can be printed quickly which is a great difference from conventional machining. Moreover, 3D printing allows a reduction in part count by printing complex sub-components as a single unit. It avoids the need to invest in costly molds until the prototypes are error-free. Damaged components can be easily replaced without the need for huge inventory giving rise to a leaner management of stocks by keeping only the raw materials needed.

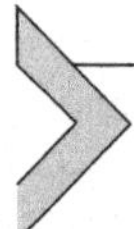

5 CURRENT GLOBAL STRATEGIC LANDSCAPE IN THE DEVELOPMENT OF ADDITIVE MANUFACTURING

5.1 Strategic Initiative of Additive Manufacturing in Different Countries

The impact of AM and its increasing adoption in many fields are as certained with the initiatives taken by governments of many countries and the booming AM businesses. With the ability to print products in-house, AM eliminates the need to keep stocks in warehouses, avoiding the long lead time and complicated logistics, bringing about a considerable transformation to the supply chain. The recognized potential of this technology that comes in the form of product design freedom, shorter lead time and reduced material wastage has attracted the attention of many industries in wanting to invest and reap these benefits. Government support through various programs to make AM technology a production mainstay is critical in driving research in this field. Leading 3D printing research centers include the Engineering and Physical Sciences Research Council (EPSRC) Center for Innovative Manufacturing in AM in the United Kingdom,

National Additive Manufacturing Innovation Institute (NAMII) Institute (subsequently known as America Makes) in the United States of America, Singapore Centre for 3D Printing (SC3DP) in Singapore and the Direct Manufacturing Research Center (DMRC) in Germany. These centers bridge the gaps between research and actual application in the industries to achieve tangible results and advance the technology.

Since July 2012, the University of Nottingham and Loughborough University have been hosting the EPSRC government funded Centre for Innovative Manufacturing in Additive Manufacturing in United Kingdom. The center has a total fund of £8.1 million from the government and numerous participating companies and is focused on the development of multimaterial and multifunctional products. This eliminates any need for assembly when multicomponent complex parts are built in a single 3D printing process and such products are widely applicable across many industries.

NAMII (or America Makes) is a potentially self-sustaining pilot institute located in Youngstown, Ohio that was established in August 2012 intended to encourage manufacturing investments in the United States. The institute aims to be a global center of excellence in 3D printing and will provide the support necessary to develop new 3D printing technologies and products. A fund of USD 70 million from the government and participating companies was initially awarded to the institute which is a consortium of 40 companies, 9 research institutes, 5 colleges, and 11 nonprofit organizations. The institute focuses on bridging research and product development gaps while educating 3D printing technologies to students, engineers, companies, and designers.

SC3DP is a nationally funded center that provides world class 3D printing facilities with close to SGD150 million funding. It is focused on creating innovative technologies and novel processes while nurturing an educated workforce and attracting talents for the growing need in AM. The four key industries that SC3DP is focusing on are: (1) Aerospace and Defense, (2) Building and Construction, (3) Marine and Offshore, and (4) Future of Manufacturing. The center serves as a one-stop shop where industries interact through projects and consultations.

Direct Manufacturing Research Center (DMRC) in Germany was founded in 2008 involving several industries and academia, namely: (1) Boeing, (2) Electro Optical Systems (EOS), (3) Evonik Industries, (4) SLM Solutions GmbH, and (5) University of Paderborn. The research center aims to build on the existing strengths and capabilities of the involved partners through the advancement of technology and equipment. The center also aims to promote technologies and equipment for companies to adopt, educate the younger generation in 3D printing, and perform market studies, benchmark processes, and scenario projections for the future in AM.

However, there is a lack of collaboration between institutes, research centers, and practitioners in the area of 3D printing. Money and time are incurred without significant contribution to the advancement of the technology. If research connections between researchers and collaborations along the whole value chain are stronger, there will be higher adoption of the technology in the industries.

5.2 Economic Landscape of Additive Manufacturing

AM techniques are developed and used to enhance productivity and competitiveness in both industrialized and developing countries. Hundreds of millions have been invested in the development and commercialization of AM in China, Singapore and several European countries. There is a wide range of application for 3D printed products. As such, it has had major impacts on numerous industries. As the challenges in AM are being addressed and eventually overcome, it is foreseen that this technology will have a significant effect on the global economic landscape. 3D printing can change the way products are fabricated as it continues to become easier to operate, more affordable and has even wider applicability, benefiting nations and industries. Some of the most significant challenges include the development of standards, improving the selection and affordability of materials, and increasing the reliability and accuracy of equipment and processes.

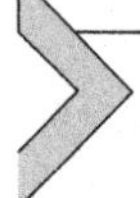

6 IMPORTANCE OF QUALITY MANAGEMENT, STANDARDS, QUALITY CONTROL, AND MEASUREMENT SCIENCES IN ADDITIVE MANUFACTURING

While many companies have explored the potential of AM for new business opportunities through novel designs that were previously impossible and will alter the makeup of supply chains, several hurdles prevent its wider adoption. One of the most critical barriers is the qualification of AM parts. Many manufacturers and users do not have utmost confidence and certainty that AM parts would exhibit consistent quality and reliability within and across different printers and geographies.

Indeed, many argue that quality assurance (QA) remains the biggest issue in AM. It is a multifaceted challenge, encompassing both the scale and scope of production. Engineers may have to relook at the entire qualification process. Certification procedures will have to be reevaluated to suit AM parts too.

In order for AM to be adopted at a wider scale, most organizations require a more sustainable and feasible approach to qualifying and certifying parts. Printer and material manufacturers seek to differentiate their products' capabilities based on their ability to print high-quality parts. However, there is a lack of a universally accepted definition of "high quality". Today, standards bodies, such as American Society for Testing and Materials (ASTM) and International Organization for Standardization (ISO) are in collaboration to develop standards for AM.

6.1 Current Challenges and Significance

Some challenges faced by AM are due to the lack of supportive frameworks for materials, software support, sustainability, and reliability. There is also a lack of standardization among AM products.

6.1.1 Materials

Metallic powders that are unsintered/unmelted in AM can often be sifted and recycled for subsequent fabrications. Liquid polymers not part of the object design, however, are discarded after every process, posing a challenge in material wastage in AM. Significant cost and energy are embedded in the creation of the initial materials used in AM and should be recycled whenever and as much as possible. Currently, a wide range of polymer resins with varying material properties is readily available, but some, like UV resins used in SLA are still toxic. This creates a safety issue on their usage. On the other hand, not all metals can be additively manufactured. Materials lacking formability contain inherent cracks due to the high thermal gradients and rapid solidification rates experienced during fabrication. Compositions of the metals should be modified in order to withstand large thermal fluctuations. Research and development of materials is one of the greatest opportunities in AM that needs to be invested in. The cost of materials used in AM processes is also significantly higher than those used in conventional processes. Nylon used in injection molding costs about USD 8 while that used in 3D printing costs about USD 80 for the same amount.

6.1.2 Software

CAD is often used to assist in the creation, modification, analysis, or optimization of designs additively manufactured. However, basic CAD software is often inadequate for the design of complex objects that are required in AM. The current CAD system is primarily designed for use with conventional manufacturing techniques where simple circles and straight lines suffice.

As such, they have limited application in exploring complex design freedoms in AM, especially in fields, such as biomimetics. Apart from its design restrictions, CAD system interfaces are usually not user-friendly and require significant user training and experience. Therefore, to fully reap the benefits of AM, these two aspects must be improved upon to propel it to its next higher peak.

6.1.3 Sustainability

Despite the ability of AM in supporting customized small-scale production with reduced material wastage, its energy-reducing economies of scale still lag in comparison with conventional manufacturing techniques. Industrial manufacturers are driven toward efficiency that results in reduced carbon footprint. With AM's ability to fabricate single or small-scale productions, efficiency can be further improved by not producing more spare stock than what are immediately required. Moreover, parallel production can be adopted to enhance the efficiency and speed of AM production processes. Industries should integrate analyses from the extraction and generation of raw materials to the manufacturing details to boost efficiency. Sustainability can be supported by AM too. One such example is, the creation of lightweight structures for aerospace applications. Energy expenditure can then be reduced and fuel savings can be achieved.

6.1.4 Reliability

Additively manufactured products, in general, lack reliability, and reproducibility as compared to the counterparts fabricated conventionally due to inherent inconsistencies in the former process. An example is the inevitable differences in arrangement and size of powder particles between the layers. The melt pool, therefore, varies in dimension between scan lines at different positions. Such random discrepancies accumulate and lower the reproducibility of additively manufactured products. Another example is the differences in recoil pressure and energy experienced by the powder at the very first instant the laser beam hits on the powder bed and when the laser beam is already scanning at a constant speed. This causes the melt pool to vary in geometry across the different scan lines, while accumulating errors as the build proceeds. Also, additively manufactured parts tend to have random poor surface finishes. Melt pools experience surface tension at the liquid–air interface, resulting in its spheroidisation and unevenness at the surface. The fluidity of the molten material is also difficult to control during fabrication, causing poor surface finish and lowering the reproducibility of parts. Also, the layer-by-layer addition of material results in microstep formations on the surfaces of the fabricated parts. The establishment of standards is, therefore, necessary and important for assurances to businesses and manufacturers that AM processes are indeed safe and reliable. ASTM Technical Committee F42 on AM technologies and ISO/TC 261 AM Technical Committee have made progress in the development of standards in AM.

7 LIABILITY, LEGALITY, AND RESPONSIBILITIES

As AM machines or 3D printers become more popular and cost-effective, the consumer and manufacturer relationship may no longer conform to the fundamentals of product liability law. Furthermore, there are also concerns on intellectual property (IP) protection. Any person can now perform a 3D scan of a product and reproduce it via AM. And how the eventual part then performs to its intended purpose is also a topic of controversy.

7.1 Liability

A significant increase in usage of 3D printers has been observed over the past decades. Despite 3D printers demonstrating the ability to produce intricate and sophisticated parts, product liability is an unavoidable implication associated with this technology. Products produced by AM could be defective for various reasons—examples include corrupted CAD files, faulty equipment, incorrect materials, or process parameters, file format deficiencies, human error in setting up the machine, human error in implementing the digital design, and so on. Given the ease and widespread availability, anyone with access to a system can be an AM producer. Considering such capability, implications arise when anyone is able to sell products made by AM. Contemporary product liability law protects consumers against manufacturers, retailers, distributors, suppliers, and others who are responsible for the defective goods [117,118]. The consumers may able to seek recourse against the commercial manufacturers if evidence is found against them.

However, with the abundance of 3D printers, the general public who may turn to a commercial printer center to print their components may lack such options of recourse. The next example demonstrates how the product liability law may not apply to such printing center. A student engaged a 3D printing center in producing a specific component. The CAD file was obtained from an online source. The student was injured while using the printed component and decided to sue the printing center to recover his loss. In this scenario, the student might not be able to successfully sue the printing center to recover his loss under the rule of strict liability. Strict liability applies when the seller is engaged in the commercial business of manufacturing and selling of products. In such cases, the printing service center is providing a service to the student and does not sell the printed products. Therefore, the printing center is not liable for the student's loss. At the most, the victim may argue that the 3D printing center did not maintain their machines properly which in turn caused the printed component to fail resulting in the injury. However, the student has to prove that the printing center has insufficiently maintained their machines. Therefore, the uncertainty of product liability is often questioned with the growth of revolutionary manufacturing and shortening of the supply chain. Manufacturers and companies should be aware of the causes and effects of liability related to AM and establish protections to prevent negative impacts against them.

7.2 Legality

Legality is one of the main concerns of AM. It is a disruptive technology that often gathers controversy on questions such as IP, product liability, regulatory, and many other areas [119,120]. Not only does AM allows the public to produce complex shapes and designs, and particularly, it also allows recreation of any existing products and possibly selling it without permission from its originator. IP protection is a major concern with the rise in popularity of AM. For example, a service center can print their own parts instead of ordering them from its vendor if they have 3D printers in their center. This move from the company may shorten the waiting time for the parts to arrive and save costs in the long run. However, these give rise to questions related to patent protection, trademarks, copyright infringement etc. Therefore, current IP protection system must evolve with the growth of AM technologies. The IP protection mechanisms should consider both physical and digital representation of the object to mitigate the impact of counterfeiting. The music industry had once faced similar problems with IP protection, and AM can learn some lessons from their history to devise and implement technological solutions to ensure IP rights of the developers are protected.

7.3 Responsibilities

Following the expiry of a patent related to FDM in 2009, the general availability of desktop 3D printers has skyrocketed. Students and hobbyists can purchase a desktop 3D printer at an affordable price for designing and producing their own 3D printed parts for either personal use or commercial purposes. Thus, this process makes the students or hobbyists both the consumer as well as the manufacturer. In such a case, the user must accept and be responsible for the outcome of the 3D printed parts. For an example, an old screw that had dropped off from a table for which a replacement could not be found. Given the obsolescence of the screw in the market, printing a new screw would be more convenient. If the new printed screw snapped causing the table to collapse, who should be the one responsible for it? There are many related questions that may arise. For example, was the correct material chosen in the first place? Was the CAD file free from errors? Although downloading open-source CAD files is easy, there are many uncertainties from a design point of view. For instance, is the file safe and reliable and does the design infringe any copyrights? With the use of 3D printers, the users should be willing to accept the associated risks and take ownership of the printed products.

8 QUESTIONS

1. What is AM?
2. What are the advantages and disadvantages of AM compared to conventional manufacturing methods?
3. What are the main categories of materials used in AM?
4. What are the main applications of AM in industries?
5. What are the current challenges of AM and their significance?

REFERENCES

[1] C.K. Chua, K.F. Leong, 3D printing and additive manufacturing: principles and applications, 5th ed., World Scientific Publishing Company, (2017).
[2] C.K. Chua, M.V. Matham, Y.J. Kim, Lasers in 3D printing and manufacturing, World Scientific Publishing Company, Singapore, (2017).
[3] C.K. Chua, W.Y. Yeong, Bioprinting: principles and applications, World Scientific Publishing Company, Singapore, (2014).
[4] J.-Y. Lee, W.S. Tan, J. An, C.K. Chua, C.Y. Tang, A.G. Fane, T.H. Chong, The potential to enhance membrane module design with 3D printing technology, J. Membr. Sci. 499 (2016) 480–490.
[5] A.T. Sutton, C.S. Kriewall, M.C. Leu, J.W. Newkirk, Powder characterisation techniques and effects of powder characteristics on part properties in powder-bed fusion processes, Virtual Phys. Prototyp. 11 (2016) 1–27.
[6] W. Wu, S.B. Tor, C.K. Chua, K.F. Leong, A. Merchant, Investigation on processing of ASTM A131 Eh36 high tensile strength steel using selective laser melting, Virtual Phys. Prototyp. 10 (2015) 187–193.

[7] C.Y.Yap, C.K. Chua, Z.L. Dong, An effective analytical model of selective laser melting, Virtual Phys. Prototyp. 11 (2016) 21–26.
[8] Y. Yang, P. Wu, X. Lin, Y. Liu, H. Bian, Y. Zhou, et al. System development, formability quality and microstructure evolution of selective laser-melted magnesium, Virtual Phys. Prototyp. 11 (2016) 173–181.
[9] K.K. Wong, J.Y. Ho, K.C. Leong, T.N. Wong, Fabrication of heat sinks by Selective Laser Melting for convective heat transfer applications, Virtual Phys. Prototyp. 11 (2016) 159–165.
[10] R. Li, Y. Shi, Z. Wang, L. Wang, J. Liu, W. Jiang, Densification behavior of gas and water atomized 316L stainless steel powder during selective laser melting, Appl. Surf. Sci. 256 (2010) 4350–4356.
[11] Z.H. Liu, C.K. Chua, K.F. Leong, K. Kempen, L. Thijs, E. Yasa, et al., A preliminary investigation on selective laser melting of M2 high speed steel, in: 5th International Conference on Advanced Research in Virtual and Rapid Prototyping, Leiria, Portugal, 2011, pp. 339–346.
[12] M. Badrossamay, T. Childs, Further studies in selective laser melting of stainless and tool steel powders, Int. J. Mach. Tools Manufacture 47 (2007) 779–784.
[13] F. Abe, K. Osakada, M. Shiomi, K. Uematsu, M. Matsumoto, The manufacturing of hard tools from metallic powders by selective laser melting, J. Mater. Proc. Technol. 111 (2001) 210–213.
[14] R. Li, J. Liu, Y. Shi, M. Du, Z. Xie, 316L stainless steel with gradient porosity fabricated by selective laser melting, J. Mater. Eng. Perform. 19 (2010) 666–671.
[15] P. Mercelis, J.-P. Kruth, Residual stresses in selective laser sintering and selective laser melting, Rapid Prototyp. J. 12 (2006) 254–265.
[16] Y.F. Shen, D.D. Gu, P. Wu, Development of porous 316L stainless steel with controllable microcellular features using selective laser melting, Mater. Sci. Technol. 24 (2008) 1501–1505.
[17] M. Wong, S. Tsopanos, C.J. Sutcliffe, I. Owen, Selective laser melting of heat transfer devices, Rapid Prototyp. J. 13 (2007) 291–297.
[18] A. Gusarov, I. Yadroitsev, P. Bertrand, I. Smurov, Heat transfer modelling and stability analysis of selective laser melting, Appl. Surf. Sci. 254 (2007) 975–979.
[19] E. Yasa, J. Deckers, J.-P. Kruth, M. Rombouts, J. Luyten, Charpy impact testing of metallic selective laser melting parts, Virtual Phys. Prototyp. 5 (2010) 89–98.
[20] K. Osakada, M. Shiomi, Flexible manufacturing of metallic products by selective laser melting of powder, Int. J. Machine Tools Manufacture 46 (2006) 1188–1193.
[21] A.B. Spierings, G. Levy, Comparison of density of stainless steel 316L parts produced with selective laser melting using different powder grades, Ann. Int. Solid Freeform Fabric. Symp. (2009) 342–353.
[22] A.B. Spierings, N. Herres, G. Levy, Influence of the particle size distribution on surface quality and mechanical properties in AM steel parts, Rapid Prototyp. J. 17 (2011) 195–202.
[23] T. Childs, C. Hauser, M. Badrossamay, Selective laser sintering (melting) of stainless and tool steel powders: experiments and modelling, Proc. Institut. Mech. Eng. Part B J. Eng. Manufacture 219 (2005) 339–357.
[24] T. Childs, C. Hauser, M. Badrossamay, Mapping and modelling single scan track formation in direct metal selective laser melting, CIRP Annal. Manufacturing Technol. 53 (2004) 191–194.
[25] K. Zeng, D. Pal, B. Stucker, A review of thermal analysis methods in Laser Sintering and Selective Laser Melting, in: Solid Freeform Fabrication Symposium, Austin, TX, USA, 2012, p. 796.
[26] K. Guan, Z. Wang, M. Gao, X. Li, X. Zeng, Effects of processing parameters on tensile properties of selective laser melted 304 stainless steel, Materials Design 50 (2013) 581–586.

[27] M. Fateri, A. Gebhardt, M. Khosravi, Numerical Investigation of selective laser melting process for 904L stainless steel, ASME 2012 Int. Mech. Eng. Congress Exposition 3 (2012) 119–124.
[28] I. Yadroitsev, P. Bertrand, B. Laget, I. Smurov, Application of laser assisted technologies for fabrication of functionally graded coatings and objects for the International Thermonuclear Experimental Reactor components, J. Nucl. Mater. 362 (2007) 189–196.
[29] Z.H. Liu, D.Q. Zhang, C.K. Chua, K.F. Leong, Crystal structure analysis of M2 high speed steel parts produced by selective laser melting, Mater. Character. 84 (2013) 72–80.
[30] K.A. Mumtaz, N. Hopkinson, Selective laser melting of thin wall parts using pulse shaping, J. Mater. Process. Technol. 210 (2010) 279–287.
[31] I. Yadroitsev, A. Gusarov, I. Yadroitsava, I. Smurov, Single track formation in selective laser melting of metal powders, J. Mater. Process. Technol. 210 (2010) 1624–1631.
[32] C.S. Wright, M. Youseffi, S.P. Akhtar, T.H.C. Childs, C. Hauser, P. Fox, Selective laser melting of prealloyed high alloy steel powder beds, Mater. Sci. Forum 514–516 (2006) 516–523.
[33] J. Milovanovic, M. Stojkovic, M. Trajanovic, Metal laser sintering for rapid tooling in application to tyre tread pattern mould, J. Sci. Ind. Res. 68 (2009) 1038.
[34] S.L. Campanelli, N. Contuzzi, A.D. Ludovico, Manufacturing of 18 Ni marage 300 steel samples by selective laser melting, Adv. Mater. Res. 83 (2010) 850–857.
[35] M. Badrossamay, E. Yasa, J. Van Vaerenbergh, J.-P. Kruth, Improving productivity rate in SLM of commercial steel powders, presented at the SME Rapid, Schaumburg, IL, USA, 2009.
[36] E. Yasa, K. Kempen, J.-P. Kruth, L. Thijs, J. Van Humbeeck, Microstructure and mechanical properties of Maraging Steel 300 after selective laser melting," in: 21st Annual International Solid Freeform Fabrication (SFF) Symposium, University of Texas, Austin, TX, USA 2010, p. 383.
[37] L. Thijs, J. Van Humbeeck, K. Kempen, E. Yasa, J.-P. Kruth, M. Rombouts, Investigation on the inclusions in maraging steel produced by Selective Laser Melting, in: 5th International Conference on Advanced Research in Virtual and Rapid Prototyping, Leiria, Portugal, 2011, p. 297.
[38] K. Kempen, E. Yasa, L. Thijs, J.-P. Kruth, J. Van Humbeeck, Microstructure and mechanical properties of selective laser melted 18Ni-300 steel, Phys Procedia 12 (2011) 255–263.
[39] C. Casavola, S. Campanelli, C. Pappalettere, Preliminary investigation on distribution of residual stress generated by the selective laser melting process, J. Strain Anal. Eng. Design 44 (2009) 93–104.
[40] L.E. Murr, E. Martinez, J. Hernandez, S. Collins, K.N. Amato, S.M. Gaytan, et al. Microstructures and properties of 17-4 PH stainless steel fabricated by selective laser melting, J. Mater. Res. Technol. 1 (2012) 167–177.
[41] M. Averyanova, E. Cicala, P. Bertrand, D. Grevey, Optimization of selective laser melting technology using design of experiments method, in: 5th International Conference on Advanced Research in Virtual and Rapid Prototyping, Leiria, Portugal, 2011, p. 459.
[42] H.K. Rafi, T.L. Starr, B.E. Stucker, A comparison of the tensile, fatigue, and fracture behavior of Ti–6Al–4V and 15-5 PH stainless steel parts made by selective laser melting, Int. J. Adv. Manufacturing Technol. 69 (2013) 1299–1309.
[43] P. Jerrard, L. Hao, K. Evans, Experimental investigation into selective laser melting of austenitic and martensitic stainless steel powder mixtures, Proc. Institut. Mech. Eng. Part B J. Eng. Manufacture 223 (2009) 1409–1416.
[44] B.-D. Joo, J.-H. Jang, J.-H. Lee, Y.-M. Son, Y.-H. Moon, Selective laser melting of Fe-Ni-Cr layer on AISI H13 tool steel, Transact. Nonferrous Metals Soc. China 19 (2009) 921–924.
[45] B. Sustarsic, S. Dolinsek, M. Jenko, V. Leskovšek, Microstructure and mechanical characteristics of DMLS tool-inserts, Mater. Manufactur Process. 24 (2009) 837–841.

[46] B. Song, S. Dong, P. Coddet, H. Liao, C. Coddet, Fabrication and microstructure characterization of selective laser-melted FeAl intermetallic parts, Surf. Coat. Technol. 206 (2012) 4704–4709.
[47] B. Song, S. Dong, H. Liao, C. Coddet, Manufacture of Fe–Al cube part with a sandwich structure by selective laser melting directly from mechanically mixed Fe and Al powders, Int. J. Adv. Manufactur. Technol. 69 (2013) 1323–1330.
[48] J.C. Walker, K.M. Berggreen, A.R. Jones, C.J. Sutcliffe, Fabrication of Fe–Cr–Al oxide dispersion strengthened PM2000 alloy using selective laser melting, Adv. Eng. Mater. 11 (2009) 541–546.
[49] A. Amanov, S. Sasaki, I.-S. Cho, Y. Suzuki, H.-J. Kim, D.-E. Kim, An investigation of the tribological and nano-scratch behaviors of Fe–Ni–Cr alloy sintered by direct metal laser sintering, Mater. Design 47 (2013) 386–394.
[50] A. Fukuda, M. Takemoto, T. Saito, S. Fujibayashi, M. Neo, D.K. Pattanayak, et al. Osteoinduction of porous Ti implants with a channel structure fabricated by selective laser melting, Acta Biomater. 7 (2011) 2327–2336.
[51] D. Gu, Y.-C. Hagedorn, W. Meiners, G. Meng, R.J.S. Batista, K. Wissenbach, et al. Densification behavior, microstructure evolution, and wear performance of selective laser melting processed commercially pure titanium, Acta Mater. 60 (2012) 3849–3860.
[52] B. Zhang, H. Liao, C. Coddet, Selective laser melting commercially pure Ti under vacuum, Vacuum 95 (2013) 25–29.
[53] H. Attar, M. Bönisch, M. Calin, L.-C. Zhang, S. Scudino, J. Eckert, Selective laser melting of in situ titanium–titanium boride composites: processing, microstructure and mechanical properties, Acta Mater 76 (2014) 13–22.
[54] K. H. Low, K.F. Leong, and C. N. Sun, Review of selective laser melting process parameters for commercially pure titanium and Ti6Al4V, in: 6th International Conference on Advanced Research in Virtual and Rapid Prototyping, Leiria, Portugal 2013, p. 71.
[55] F. Abe, E.C. Santos, Y. Kitamura, K. Osakada, M. Shiomi, Influence of forming conditions on the titanium model in rapid prototyping with the selective laser melting process, Proc. Institut. Mech. Eng. Part C J. Mech. Eng. Sci. 217 (2003) 119–126.
[56] E. Santos, K. Osakada, M. Shiomi, Y. Kitamura, F. Abe, Microstructure and mechanical properties of pure titanium models fabricated by selective laser melting, Proc. Institut. Mech. Eng. Part C J. Mech. Eng. Sci. 218 (2004) 711–719.
[57] E. Santos, K. Osakada, M. Shiomi, M. Morita, F. Abe, Fabrication of titanium dental implants by selective laser melting, in: Fifth International Symposium on Laser Precision Microfabrication, Nara, Japan 2004, pp. 268–273.
[58] A. Barbas, A.-S. Bonnet, P. Lipinski, R. Pesci, G. Dubois, Development and mechanical characterization of porous titanium bone substitutes, J. Mech. Behavior Biomed. Mater. 9 (2012) 34–44.
[59] M. Simonelli, Y.Y. Tse, C. Tuck, Effect of the build orientation on the mechanical properties and fracture modes of SLM Ti-6Al-4V, Mater. Sci. Eng. A 616 (2014) 1–11.
[60] B. Vrancken, L. Thijs, J.-P. Kruth, J. Van Humbeeck, Heat treatment of Ti6Al4V produced by Selective Laser Melting: Microstructure and mechanical properties, J. Alloys Comp. 541 (2012) 177–185.
[61] B. Song, S. Dong, B. Zhang, H. Liao, C. Coddet, Effects of processing parameters on microstructure and mechanical property of selective laser melted Ti6Al4V, Mater. Design 35 (2012) 120–125.
[62] B. Vandenbroucke, J.-P. Kruth, Selective laser melting of biocompatible metals for rapid manufacturing of medical parts, Rapid Prototyp. J. 13 (2007) 196–203.
[63] T. Marcu, M. Todea, I. Gligor, P. Berce, C. Popa, Effect of surface conditioning on the flowability of Ti6Al7Nb powder for selective laser melting applications, Appl. Surf. Sci. 258 (2012) 3276–3282.

[64] E. Chlebus, B. Kuźnicka, T. Kurzynowski, B. Dybała, Microstructure and mechanical behaviour of Ti–6Al–7Nb alloy produced by selective laser melting, Mater. Character. 62 (2011) 488–495.
[65] T. Sercombe, N. Jones, R. Day, A. Kop, Heat treatment of Ti-6Al-7Nb components produced by selective laser melting, Rapid Prototyp. J. 14 (2008) 300–304.
[66] L.C. Zhang, D. Klemm, J. Eckert, Y.L. Hao, T.B. Sercombe, Manufacture by selective laser melting and mechanical behavior of a biomedical Ti–24Nb–4Zr–8Sn alloy, Scripta Mater. 65 (2011) 21–24.
[67] A. Zieliński, S. Sobieszczyk, W. Serbiński, T. Seramak, A. Ossowska, Materials design for the titanium scaffold based implant, Solid State Phenom. 183 (2012) 225–232.
[68] M. Speirs, J.V. Humbeeck, J. Schrooten, J. Luyten, J.-P. Kruth, The effect of pore geometry on the mechanical properties of selective laser melted Ti-13Nb-13Zr scaffolds, Procedia CIRP 5 (2013) 79–82.
[69] K. Kempen, L. Thijs, J. Van Humbeeck, J.-P. Kruth, Mechanical properties of AlSi10Mg produced by selective laser melting, Phys. Procedia 39 (2012) 439–446.
[70] K. Mumtaz, N. Hopkinson, Top surface and side roughness of Inconel 625 parts processed using selective laser melting, Rapid Prototyp. J. 15 (2009) 96–103.
[71] C. Paul, S. Mishra, C. Premsingh, P. Bhargava, P. Tiwari, L. Kukreja, Studies on laser rapid manufacturing of cross-thin-walled porous structures of Inconel 625, Int. J. Adv. Manufact. Technol. 61 (2012) 757–770.
[72] I. Yadroitsev, L. Thivillon, P. Bertrand, I. Smurov, Strategy of manufacturing components with designed internal structure by selective laser melting of metallic powder, Appl. Surf. Sci. 254 (2007) 980–983.
[73] K. Amato, S. Gaytan, L. Murr, E. Martinez, P. Shindo, J. Hernandez, et al. Microstructures and mechanical behavior of Inconel 718 fabricated by selective laser melting, Acta Mater. 60 (2012) 2229–2239.
[74] Z. Wang, K. Guan, M. Gao, X. Li, X. Chen, X. Zeng, The microstructure and mechanical properties of deposited-IN718 by selective laser melting, J. Alloys Comp. 513 (2012) 518–523.
[75] F. Wang, Mechanical property study on rapid additive layer manufacture Hastelloy® X alloy by selective laser melting technology, Int. J. Adv. Manufact. Technol. 58 (2012) 545–551.
[76] F. Wang, X.H. Wu, D. Clark, On direct laser deposited Hastelloy X: dimension, surface finish, microstructure and mechanical properties, Mater. Sci. Technol. 27 (2011) 344–356.
[77] D. Tomus, T. Jarvis, X. Wu, J. Mei, P. Rometsch, E. Herny, et al. Controlling the microstructure of Hastelloy-X components manufactured by selective laser melting, Phys. Procedia 41 (2013) 823–827.
[78] T. Vilaro, C. Colin, J.-D. Bartout, L. Nazé, M. Sennour, Microstructural and mechanical approaches of the selective laser melting process applied to a nickel-base superalloy, Mater. Sci. Eng. 534 (2012) 446–451.
[79] E.O. Olakanmi, Selective laser sintering/melting (SLS/SLM) of pure Al, Al–Mg, and Al–Si powders: Effect of processing conditions and powder properties, J. Mater. Process. Technol. 213 (2013) 1387–1405.
[80] E. Brandl, U. Heckenberger, V. Holzinger, D. Buchbinder, Additive manufactured AlSi10Mg samples using Selective Laser Melting (SLM): Microstructure, high cycle fatigue, and fracture behavior, Mater. Design 34 (2012) 159–169.
[81] E. Louvis, P. Fox, C.J. Sutcliffe, Selective laser melting of aluminium components, J. Mater. Process. Technol. 211 (2011) 275–284.
[82] K. Kempen, L. Thijs, E. Yasa, M. Badrossamay, W. Verheecke, J.-P. Kruth, Process optimization and microstructural analysis for selective laser melting of AlSi10Mg, in: Solid Freeform Fabrication Symposium, Austin, TX, USA 2011.

[83] L.E. Loh, Z.H. Liu, D.Q. Zhang, M. Mapar, S.L. Sing, C.K. Chua, et al. Selective laser melting of aluminium alloy using a uniform beam profile, Virtual Phys. Prototyp. 9 (2014) 11–16.
[84] L. Thijs, K. Kempen, J.-P. Kruth, J. Van Humbeeck, Fine-structured aluminium products with controllable texture by selective laser melting of pre-alloyed AlSi10Mg powder, Acta Mater. 61 (2013) 1809–1819.
[85] F. Calignano, D. Manfredi, E. Ambrosio, L. Iuliano, P. Fino, Influence of process parameters on surface roughness of aluminum parts produced by DMLS, Int. J. Adv. Manufactur. Technol. 67 (2013) 2743–2751.
[86] D. Buchbinder, H. Schleifenbaum, S. Heidrich, W. Meiners, J. Bültmann, High power selective laser melting (HP SLM) of aluminum parts, Phys. Procedia 12 (2011) 271–278.
[87] M. Ameli, B. Agnew, P.S. Leung, B. Ng, C. Sutcliffe, J. Singh, et al. A novel method for manufacturing sintered aluminium heat pipes (SAHP), Appl. Thermal Eng. 52 (2013) 498–504.
[88] K.G. Prashanth, S. Scudino, H.J. Klauss, K.B. Surreddi, L. Loeber, Z. Wang, et al. Microstructure and mechanical properties of Al–12Si produced by selective laser melting: Effect of heat treatment, Mater. Sci. Eng. A 590 (2014) 153–160.
[89] X.J. Wang, L.C. Zhang, M.H. Fang, T.B. Sercombe, The effect of atmosphere on the structure and properties of a selective laser melted Al–12Si alloy, Mater. Sci. Eng. 597 (2014) 370–375.
[90] Y. Tang, H.T. Loh, Y.S. Wong, J.Y.H. Fuh, L. Lu, X. Wang, Direct laser sintering of a copper-based alloy for creating three-dimensional metal parts, J. Mater. Process. Technol. 140 (2003) 368–372.
[91] Z.H. Liu, D.Q. Zhang, S.L. Sing, C.K. Chua, L.E. Loh, Interfacial characterisation of SLM parts in multi material processing: metallurgical diffusion between 316L stainless steel and C18400 copper alloy, Mater. Character. 94 (2014) 116–125.
[92] D.Q. Zhang, Z.H. Liu, C.K. Chua, Investigation on forming process of copper alloys via selective laser melting, in: Proceedings of the 6th International Conference on Advanced Research in Virtual and Rapid Prototyping, Leiria, Portugal, 2013, p. 285.
[93] R. Li, Y. Shi, J. Liu, Z. Xie, Z. Wang, Selective laser melting W–10 wt. % Cu composite powders, Int. J. Adv. Manufactur. Technol. 48 (2010) 597–605.
[94] D. Zhang, Q. Cai, J. Liu, R. Li, Research on process and microstructure formation of W-Ni-Fe alloy fabricated by selective laser melting, J. Mater. Eng. Perform. 20 (2011) 1049–1054.
[95] M.M. Savalani, C.C. Ng, H.C. Man, Selective laser melting of magnesium for future applications in medicine," in: 2010 International Conference on Manufacturing Automation, Hong Kong, 2010, pp. 50–54.
[96] M., Giesekel, C. Noelkel, S. Kaierlel, V. Weslingl, H. Haferkarnp, Selective laser melting of magnesium and magnesium alloys, in: N. Hort, S.N. Mathaudhu, N.R. Neelameggham, M. Alderman (Eds.), Magnesium Technology 2013, pp. 65–68.
[97] B. Zhang, H. Liao, C. Coddet, Effects of processing parameters on properties of selective laser melting Mg–9% Al powder mixture, Mater. Design 34 (2012) 753–758.
[98] L. Wu, H. Zhu, X. Gai, Y. Wang, Evaluation of the mechanical properties and porcelain bond strength of cobalt-chromium dental alloy fabricated by selective laser melting, J. Prosthetic Dentist. 111 (2014) 51–55.
[99] X.-z. Xin, J. Chen, N. Xiang, B. Wei, Surface properties and corrosion behavior of Co–Cr alloy fabricated with selective laser melting technique, Cell Biochem. Biophys. 67 (2013) 983–990.
[100] D. Zhang, Q. Cai, J. Liu, Formation of nanocrystalline tungsten by selective laser melting of tungsten powder, Mater. Manufactur. Process. 27 (2012) 1267–1270.
[101] K. Deprez, S. Vandenberghe, K. Van Audenhaege, J. Van Vaerenbergh, R. Van Holen, Rapid additive manufacturing of MR compatible multipinhole collimators with selective laser melting of tungsten powder, Med. Phys. 40 (2013) 012501.

[102] D.Q. Zhang, Q.Z. Cai, J.H. Liu, L. Zhang, R.D. Li, Select laser melting of W–Ni–Fe powders: simulation and experimental study, Int. J. Adv. Manufactur. Technol. 51 (2010) 649–658.
[103] M. Khan, P. Dickens, Selective laser melting (SLM) of gold (Au), Rapid Prototyp. J. 18 (2012) 81–94.
[104] J. Jhabvala, E. Boillat, T. Antignac, R. Glardon, On the effect of scanning strategies in the selective laser melting process, Virtual Phys. Prototyp. 5 (2010) 99–109.
[105] J.J. Brandner, E. Hansjosten, E. Anurjew, W. Pfleging, K. Schubert, Microstructure devices generation by selective laser melting, in: Lasers and Applications in Science and Engineering, Bellingham, WA, USA, 2007.
[106] I. Shishkovsky, I. Yadroitsev, P. Bertrand, I. Smurov, Alumina–zirconium ceramics synthesis by selective laser sintering/melting, Appl. Surf. Sci. 254 (2007) 966–970.
[107] H. Yves-Christian, W. Jan, M. Wilhelm, W. Konrad, P. Reinhart, Net shaped high performance oxide ceramic parts by selective laser melting, Phys. Procedia 5 (2010) 587–594.
[108] J. Wilkes, Y.-C. Hagedorn, W. Meiners, K. Wissenbach, Additive manufacturing of ZrO2-Al2O3 ceramic components by selective laser melting, Rapid Prototyp. J. 19 (2013) 51–57.
[109] P. Regenfuss, A. Streek, F. Ullmann, C. Kühn, L. Hartwig, M. Horn, et al. Laser micro sintering of ceramic materials, part 1, Interceram 56 (2007) 420–422.
[110] X.H. Wang, J.Y.H. Fuh, Y.S. Wong, Y.X. Tang, Laser sintering of silica sand—mechanism and application to sand casting mould, Int. J. Adv. Manufactur. Technol. 21 (2003) 1015–1020.
[111] F.-H. Liu, Synthesis of bioceramic scaffolds for bone tissue engineering by rapid prototyping technique, J. Sol-Gel Sci. Technol. 64 (2012) 704–710.
[112] C.Y. Yap, C.K. Chua, Z. Dong, Z.H. Liu, D.Q. Zhang, Single track and single layer melting of silica by Selective Laser Melting, in: 6th International Conference on Advanced Research in Virtual and Rapid Prototyping, Leiria, Portugal, 2013, p. 261.
[113] P. Regenfuss, A. Streek, F. Ullmann, C. Kühn, L. Hartwig, M. Horn, et al. Laser micro sintering of ceramic materials, part 2, Interceram 57 (2008) 6–9.
[114] J. Wilkes, K. Wissenbach, Rapid manufacturing of ceramic components for medical and technical applications via selective laser melting, Proceedings of Euro-uRapid, 2006, A4/1.
[115] G. Manob, Processing and characterization of lithium aluminosilicate glass parts fabricated by selective laser melting, Master of Engineering, Department of Mechanical Engineering, National University of Singapore, Singapore, 2004.
[116] G. Manob, L. Lu, J.Y.H. Fuh, Y.B. Cheng, Selective laser melting of Li_2O. Al_2O_3. SiO_2 (LAS) glass powders, Mater. Sci. Forum 437–438 (2003) 249–252.
[117] A. Harris, The effects of in-home 3D printing on product liability law, J. Sci. Policy Gov. 6 (2015) http://www.sciencepolicyjournal.org/uploads/5/4/3/4/5434385/harris_new_ta1_1.2.2015_lb_mg.pdf.
[118] M.M. Eckstein A.T. Brown, 3D printing and its uncertain products liability landscape (2016). Available from: http://www.industryweek.com/emerging-technologies/3d-printing-and-its-uncertain-products-liability-landscape
[119] M.M. Eckstein, 3D printing raises new legal questions (2016). Available from: http://www.industryweek.com/intellectual-property/3d-printing-raises-new-legal-questions
[120] Y. Lakhdar. (2016, 26/12/2016). *Additive manufacturing and intellectual property protection: an overview*. Available: https://www.linkedin.com/pulse/additive-manufacturing-intellectual-property-yazid-lakhdar.

CHAPTER TWO

Roadmap on Additive Manufacturing Standards

Contents

Standards, Quality Control, and Measurement Sciences in 3D Printing and Additive Manufacturing
http://dx.doi.org/10.1016/B978-0-12-813489-4.00002-7

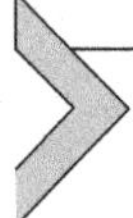

1 INTRODUCTION TO STANDARDS FOR ADDITIVE MANUFACTURING

Standards are formal documents developed by standards organizations, such as International Organization for Standardization (ISO), American Society for Testing and Materials (ASTM), German Institute for Standardization (DIN) and so on, in conjunction with relevant industry partners, to set, validate, and certify technical and safety requirements [1]. Standards satisfy the growing needs of different industries, such as consumer, trade, and industry sectors, through credible verification of a product or market performance [2]. In the context of AM, it is widely recognized that the lack of AM standards resulted in the slow adoption of AM systems into industry processes [3].

Although there is a broad spectrum of standards that are in place for conventional manufacturing practices (casting, extrusion, machining, injection molding, etc.), they are not suitable for AM applications due to several factors. In AM, components are manufactured layer-wise [4], which results in anisotropic properties throughout the component. A component fabricated by AM, when compared to a forged component, exhibits different microstructures and mechanical properties [5]. Similarly, the surface finish of an AM part without any postprocessing is coarser than a machined or forged part [6]. Additionally, AM processes affect the microstructure, mechanical properties and finishing of a part. For example, a metal component fabricated by SLM possesses different properties from a similar metallic component fabricated by EBM [7]. Without standards in place, it would not be possible to conduct a proper comparison across and within different AM processes.

Therefore, standardization is important to the AM industry. The lack of standards has resulted in slow adoption of AM technology, particularly in industries that require certification, such as aerospace, medical, automotive, and so on. This is currently being addressed by standards bodies throughout the world, such as ASTM that is working closely with industry partners to develop and maintain a set of common standards for AM.

2 IMPORTANCE OF STANDARDIZATION

AM industry is a rapidly developing market, with high growth rates since the introduction of the first commercial AM machine, SLA-1, by 3D Systems in 1987 [8]. Wohlers Report 2016 reported an impressive compounded annual growth rate of 26.2% over the past 27 years for the AM industry [9]. It was reported that the aerospace sector has grown by about 4.3% since 2013, and sectors like academic institutions and government/military have grown larger too (Fig. 2.1 describes various sectors using AM technology). Industries, ranging from industrial machines, consumer products, and electronics, automotive, aerospace, and medical, have shown an increasing adoption rate of AM technology. This increased adoption is due to the ability of AM to manufacture parts that are functional and with a higher turnover. Customers are able to obtain prototypes faster through AM, at a lower cost, as compared to traditional rapid prototyping (RP).

The introduction of AM will reduce space consumption and processing time. Complex parts can be made with fewer processes and higher turnover rate. However, even though it is possible to produce these parts faster, in some applications certifications are required to ensure that AM parts are compatible, reliable and safe to use. Craig Barrett, a former CEO of Intel, gave an example of why standards are important. He mentioned that a movie made in China must be able to play on a player that is delivered to the USA, and similarly, a movie made in the USA must be able to be played

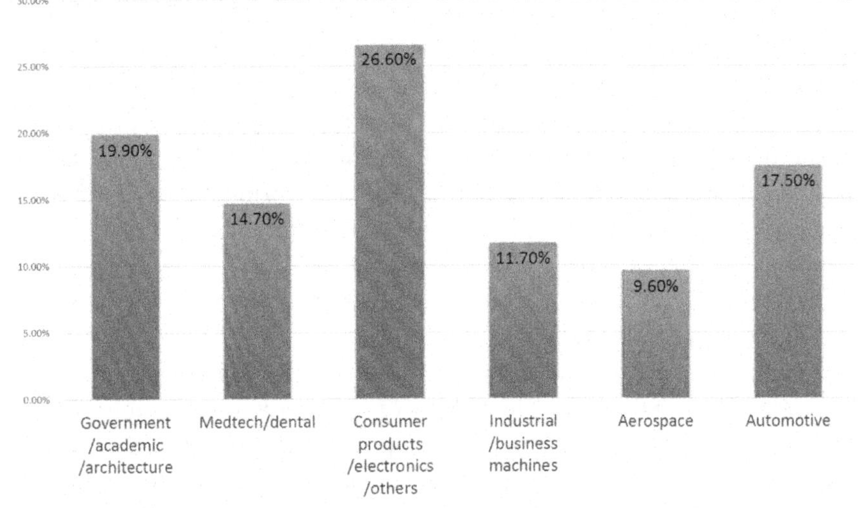

Figure 2.1 ***List of Sectors Using AM Technology [9].***

in China [1]. This can only be achieved through standardization. Therefore, AM systems must follow standards to ensure reliability and compatibility throughout the industry.

High technology, value-added, and competitive industries, such as aerospace and medical industries, demand complex, high performance, and precise parts [3]. Not only do AM parts require proper certification before it can be introduced into their components and systems, the parts have to match performance standards of conventional machined parts too. Highly regulated industries like aerospace and medical developed in-house tests to evaluate the performance level of their AM components for assurance. However, without a standard to benchmark against, it is challenging to ensure that the AM parts are "fit for use" and meet the mechanical and dimensional requirements coupled with specific quality assurance and test methods to use.

An extensive number of tests are conducted by AM companies to provide information and material datasheets to their customers. Companies develop their own tests in order to satisfy the needs of their customers but not all tests conducted refer to an available standard due to limited applicability. Tests that are not referenced to any standards will not have any form of certification to verify that an AM part has passed quality and performance tests. The lack of standards for AM is, therefore a barrier that needs to be overcome before AM can fully integrate with the industry.

The development of standards for AM is sluggish as it is affected by budget and time constraints [10]. AM has a history of about 30 years [8], but till date, the defacto industry standards are works of companies that happen to be widely used and accepted by the AM community. For example, the industry defacto standard file format for most AM machines is STereoLithography File (STL), which was developed by 3D Systems [11]. STL has been used as the norm for AM systems for over 2 decades as there was no defined standard then. Other file formats that have been developed are STEP, IGL, IGES, and so on [12]. These file formats have their own advantages and disadvantages, however, not all AM machines are able to process them [13]. In order to standardize the file format for ease of usage, ISO and ASTM established a new standard, ISO/ASTM 52915-13 [14], to replace the aging STL file format.

The updates of standards are time-consuming due to the fact that standards are developed on a voluntary basis. Standards organisations (ISO, ASTM, DIN, etc.) have begun developing new standards from best practices in the industry. From a survey conducted by J. Munguia, it was found that about 50% of the participants find current standards (ISO, ASTM, DIN, etc.) not applicable, and 37% would apply the standards partially to their work [12].

3 HISTORY OF FORMATION OF COMMITTEE FOR STANDARDS

Prior to the introduction of AM standards, the industry used defacto standards and best practices developed through testing and experimentation for AM processes. The Society of Manufacturing Engineers established the Rapid Prototyping Association in response to aid the growing AM industry [15]. ASTM formed a subcommittee E28.16 in 1999 to create new standards for mechanical testing of AM components [16]. E28.16 was the first subcommittee to evaluate the performances of AM parts. However, there were no official bodies during that time that took on the task to develop and establish standards for the AM industry.

The F42 committee, formed by ASTM in 2009, is the first official standards body for AM [17]. The committee meets twice a year to develop and publish standards in the annual book of ASTM standards, volume 10.04 for the AM industry. The next committee was established by ISO in 2011, under ISO/TC 261, whose scope is to standardize AM fundamentals [18]. In Europe, CEN/TC438 was established in 2015 to address the needs of standards in the European Union [19].

Presently, there are more than 400 members representing 23 different countries in the F42 committee. The committee has developed and approved 13 standards listed in Table 2.1, and two of which are jointly developed with ISO/TC 261 [20]. The committee is further divided into eight technical subcommittees (TC) that take charge of developing standards for different aspects of AM. The subcommittees are as follows:

- F42.01 Test Methods
- F42.04 Design
- F42.05 Material and Processes
- F42.90 Executive
- F42.91 Terminology
- F42.94 Strategic Planning
- F42.95 US TAG to ISO TC 261

ASTM also has a list of on going work items that are in development as shown in Table 2.2.

New work items are new standards or a revision of existing standard that is under development by a committee [17]. They are published by ASTM

Table 2.1 List of AM technology standards developed by ASTM and ISO [21]

Subcommittee	Standards published
F42.04 Design	ISO/ASTM52915–16: standard specification for additive manufacturing file format (AMF) version 1.2
F42.05 Material and Processes	F2924–14: standard specification for additive manufacturing titanium-6 aluminum-4 vanadium with powder bed fusion
	F3001–14: standard specification for additive manufacturing titanium-6 aluminum-4 vanadium ELI (Extra Low Interstitial) with powder bed fusion
	F3049–14: standard guide for characterizing properties of metal powders used for additive manufacturing processes
	F3055–14a: standard specification for additive manufacturing nickel alloy (UNS N07718) with powder bed fusion
	F3056–14e1: standard specification for additive manufacturing nickel alloy (UNS N06625) with powder bed fusion
	F3091/F3091M–14: standard specification for powder bed fusion of plastic materials
	F3184–16: standard specification for additive manufacturing stainless steel alloy (UNS S31603) with powder bed fusion
	F3187–16: standard guide for directed energy deposition of metals
F42.91 Terminology	ISO/ASTM52900–15: standard terminology for additive manufacturing—general principles—terminology
F42.01 Test Methods	F2971–13: standard practice for reporting data for test specimens prepared by additive manufacturing
	F3122–14: guide for evaluating mechanical properties of metal materials made via additive manufacturing processes
	ISO/ASTM52921–13: standard terminology for additive manufacturing-coordinate systems and test methodologies

for interested stakeholders to provide suggestions, even if they are not part of the committee. It is notable that standards are living documents, which will change as and when needed for reviews, to address any changes that happen in the AM community.

ISO/TC 261, which was established by ISO in 2011, consists of 20 participating countries and five observing countries. They have a total of 16 subcommittees, comprising four technical subcommittees (working group) that focus on the development of standards for AM [18]. The four subcommittees are:

- ISO/TC 261/WG 1 Terminology
- ISO/TC 261/WG 2 Methods, processes and materials
- ISO/TC 261/WG 3 Test methods
- ISO/TC 261/WG 4 Data and Design

Table 2.2 List of ASTM AM work items [21]

Subcommittee	Work items
F42.01 Test Methods	WK56649: standard practice/guide for intentionally seeding flaws in additively manufactured (AM) parts WK49229: orientation and location dependence mechanical properties for metal additive manufacturing WK55297: additive manufacturing – General principles – standard test artifacts for additive manufacturing WK55610: the characterization of powder flow properties for additive manufacturing applications
F42.04 Design	WK38342: new guide for design for additive manufacturing WK48549: new specification for AMF support for solid modeling: voxel information, constructive solid geometry representations and solid texturing WK51841: principles of design rules in additive manufacturing
F42.05 Materials and Processes	WK51282: additive manufacturing, general principles, requirements for purchased AM parts WK51329: new specification for additive manufacturing cobalt-28 chromium-6 molybdenum alloy (UNS R30075) with powder bed fusion WK37654: new guide for directed energy deposition of metals WK48732: new specification for additive manufacturing stainless steel alloy (UNS S31603) with powder bed fusion WK53423: additive manufacturing alsi10 mg with powder bed fusion WK53425: thermal post processing of metal powder bed fusion parts WK53878: additive manufacturing—material extrusion based additive manufacturing of plastic materials—part 1: Feedstock materials WK53879: additive manufacturing—material extrusion based additive manufacturing of plastic materials—part 2: Process-equipment WK53880: additive manufacturing—material extrusion based additive manufacturing of plastic materials: final part specification

Under the four technical subcommittees are 9 joint groups (JG) and 1 ad hoc group (AH) made up of members from ISO/TC 261 and ASTM F42. The 9 joint groups and ad hoc group are categorized into different technical subcommittees as illustrated in Fig. 2.2.

Under the committee, a total of 5 ISO standards have been published (Table 2.3). Two of these standards are collaborations between ISO and ASTM under the Partner Standards Development Organization (PSDO) agreement.

Figure 2.2 ***Structure of ISO/TC261 [22].***

Table 2.3 Standards and projects under the direct responsibility of ISO/TC 261 Secretariat and its SCs [23]

Standards published
ISO 17296-2:2015: additive manufacturing—general principles—part 2: overview of process categories and feedstock
ISO 17296-3:2014: additive manufacturing—general principles—part 3: main characteristics and corresponding test methods
ISO 17296-4:2014: additive manufacturing—general principles—part 4: overview of data processing
ISO/ASTM 52915:2016: atandard specification for additive manufacturing file format (AMF) Version 1.2
ISO/ASTM 52921:2013: atandard terminology for additive manufacturing—coordinate systems and test methodologies
ISO/ASTM 52900:2015: additive manufacturing: general principles—terminology

In Europe, European Committee for Standardization (CEN) CEN/TC 438 was formed to address the need for AM standards due to the rapid growth of AM technology in the industry. Established in 2015 as a result of findings by Support Action for Standardization in AM (SASAM), the scope of the committee is to provide a complete set of EU standards, and adopt the Vienna agreement with ISO/TC 261 for consistency [24]. The main objectives of CEN/TC 438 are [19]:

- To provide a complete set of European standards on processes, test procedures, quality parameters, supply agreements, fundamentals and vocabulary based, as far as possible on international standardization work. The aim is to apply the Vienna agreement with ISO/TC 261 "Additive Manufacturing" (DIN) to ensure consistency and harmonization.
- To strengthen the link between European research programs and standardization in additive manufacturing.
- To ensure visibility in the European standardization in AM by centralizing standardization initiatives in Europe.

CEN/TC 438 has not published any standards as of 2015. Instead, they chose to adopt ISO standards for usage in EU. National standards bodies from different countries in the EU (AFNOR, AITA, etc.) have established mirror committees with references to CEN/TC 438 and ISO/TC 261 for AM standards.

National and international associations in China, Japan, Korea, and Singapore have also established similar mirror committees that align to ISO/TC 261 [3]. ISO, ASTM and other standardizing bodies organized meetings in various worldwide conventions to establish and delegate task in developing standards, through roadmaps, for the AM industry.

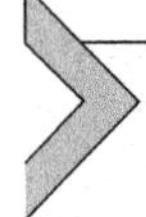

4 WORK PLAN AND ROADMAP OF JOINT COMMITTEE BETWEEN ASTM, ISO, AND THROUGHOUT THE WORLD

Roadmaps for AM have been in place for almost two decades. A pioneer roadmap was developed by the United States Department of Energy (DOE) in 1994, focusing on 3 areas of advanced rapid manufacturing of which one was on RP (or AM) [25]. In 1997, the National Institute of Standards and Technology (NIST) engaged the industry through a workshop for issues pertaining to AM. The workshop, entitled "Measurement and standards issues in rapid prototyping", targeted specific metrology issues and standards requirement by the RP community, which will be further discussed in Chapter 3. The following year, National Center for Manufacturing Science (NCMS) narrowed down the focus into building a roadmap for AM, based on the roadmap established in 1994. Despite the efforts, standards development was not a main focus of the roadmap. Instead, more efforts were put into pushing AM technology into the industry.

A decade later, National Science Foundation (NSF) and Office of Naval Research hosted a workshop to establish a new roadmap for AM. This roadmap shifted the focus from the industry to research and academia. The workshop highlighted the need to develop and adopt internationally recognized standards [25]. As a result, more focus was placed on standards and subsequent work plans and workshops in the next few years were hosted by standards organizations for establishing roadmaps for standards in AM.

The standards for AM were developed by its own standards bodies independently, prior to any form of collaboration. However, independent works from different bodies resulted in duplicate standards. To rectify the situation, ASTM and ISO/TC 261, under the jurisdiction of the PSDO agreement, came to a cooperation agreement to jointly develop international standards for AM. The PSDO agreement was approved at an ISO council meeting, held at New Delhi, India, on September 2011 by ASTM President James Thomas and ISO Secretary-General Rob Steele [26,27]. The PSDO agreement [28] covers:

- fast tracking the adoption process of an ASTM international standard as an ISO final draft international standard,
- formal adoption of a published ISO standard by ASTM international,
- maintenance of published standards, and
- publication, copyright, and commercial arrangements.

This agreement allows both organizations to adopt and jointly develop AM standards for use internationally. At the same time, PSDO agreement maximizes resource by eliminating duplicate standards, optimizing manpower, reducing the downtime of standards development, and increasing publication rates for the AM industry [29]. The first two approved standards under the PSDO agreement are ISO/ASTM 52921:2013 and ISO/ASTM 52915:2013. With the PSDO agreement, it will lead to worldwide and consolidated standards [28].

In 2013, two planning sessions were conducted by ASTM International and ISO. The first planning session was in Philadelphia (USA) and the latter in Nottingham (UK) [30]. Both meetings involved members from ASTM F42 and ISO/TC261 on the development of AM standards [20]. The result of the meetings was a joint plan for AM standards development, which will be reviewed and updated on a regular basis.

The objectives of the joint plan are:

- Bringing AM industry experts together from ISO/TC 261 and ASTM F42.
- Identifying specific standards needs common to the AM industry.
- Aligning standards roadmaps, resulting in a joint roadmap common to ISO/TC 261 and ASTM F42 interests.

- Determining how the two groups can best work together.
- Determining the priorities for specific AM standards.

Both ASTM F42 and ISO/TC 261 committees review existing standards, roadmap documents, and proposals to align their interests. Standards are grouped into key categories to achieve a common structure for easy reference. The approved structure consists of three levels, which are:

- *General standards*: standards that specify general concepts, common requirements, or are generally applicable to most type of AM materials, processes, and applications
- *Category standards*: standards that specify requirements that are specific to a material category, or process category
- *Specialized standards*: standards that specify requirements that are specific to a material, process, or application

Fig. 2.3 illustrates the agreed-upon common structure of AM standards.

EU expressed their interests in 2014 in the development of AM standards. Funded by the European Commission, SASAM began to coordinate and integrate standardization activities for EU to accelerate the growth of AM industrial processes [31]. An 18-month project, SASAM seeks to create a roadmap for standardization of AM technology, and address immediate needs of the industry and prepares for long-term development [32]. ISO/TC261, ASTM F42, and CEN/TC 438 collaborated and generated a roadmap report, to assist in developing an industrial standard for industrial application and stimulating innovation for the AM industry.

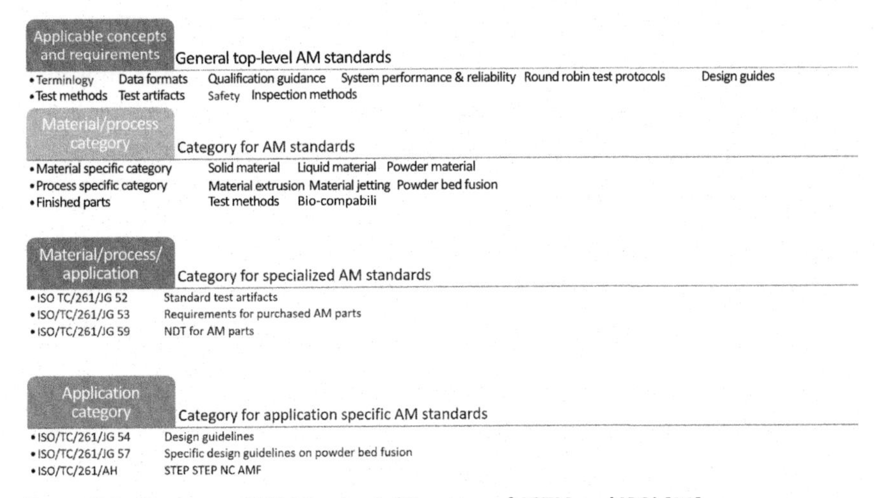

Figure 2.3 ***Structure of AM Standards (Courtesy of ASTM and ISO) [20].***

The roadmap was constructed from three main tasks:

- Gather and evaluate information from other relevant roadmaps and most important developments within this sector.
- Transform the outcome and conclusions from this information gathering and match this with the AM stakeholders and requirements survey.
- Gather feedback, finalize, and publish the roadmap for standardization serving the additive manufacturing technology.

The SASAM project also investigated the inputs of the European AM standards stakeholders through a survey, which consisted of a total 102 respondents from industry, research, and government bodies [31].

The findings from the survey include:

- Standards are in use by a majority of the participants.
- There is an urgent need for AM standards.
- It is of great importance that the AM standards will be globally and internationally accepted.
- The requirements of the customer (end user of AM parts) are main drivers for using standards; also for upcoming AM standards.
- The prioritized topics for AM standardization are materials, processes/methods and test methods.
- The reliability of machines and processes is expected to improve as a result of the development and application of standards.
- The most common argument for the need of standards is quality or qualification (system qualification, material quality, part quality, and quality control).
- Market opportunities are directly related to future standards.

SASAM highlighted the needs of EU stakeholders in the development of AM standards. This standardization activity will allow the AM industry in EU to rapidly expand into existing AM business (aerospace and medical) and new sectors. After evaluating existing documents, and joint work plan between ASTM F42 and ISO/TC 261, key agreements were drafted based on the following principles [32,33]:

- One set of AM standards—to be used all over the world.
- Common roadmap and organizational structure for AM standards.
- Use and elaborate upon existing standards, modified for AM when necessary.
- ISO/TC 261 and ASTM F42 should work together and in the same direction for efficiency and effectiveness.

A consensus was reached by all parties to follow the guidelines drafted out by SASAM. Fig. 2.4 shows a similar roadmap published by SASAM in 2015.

TRL	Goals	
NIL	Certification for quality of life enhancing applications	2014–2018
NIL	Certification for energy saving applications	2016–2020
NIL	General mechanical applications	2018–2022
	Productivity/other	
5-6	Post-processing	2016–2018
7-9	Process monitoring	2016–2018
5-6	Lattice Structures	2018–2022
5-6	Database with material properties	2015–2019
	Materials	
5-6	Ti grade 1	2014–2016
5-6	CoCr	2017–2019
1-4	Al	2018–2020
5-6	TAl	2015–2017
5-6	Tool steel	2018–2020
7-9	TAl 64	2016–2018
5-6	Inconel 635 and 718	2016–2017
5-6	Stainless steel	2018–2020
1-4	Gold and bronze	2021–2022
7-9	PA12	2015–2017
1-4	ABS	2017–2019
7-9	PA11	2016–2018
5-6	MED610	2019–2022
5-6	Rubber-like	2018–2020
5-6	PEEK	2020–2022
7-9	PA flame retardant	2016–2018
1-4	Ceramics alumina	2019–2022
	Process stability/product quality	
1-4	Fatigue testing	2015–2017
1-4	Creep	2019–2021
7-9	Geometrical tolerance	2014–2015
5-6	Flexural strength	2016–2018
5-6	Shear resistance	2020–2022
5-6	Impact strength	2017–2019
7-9	Surface texture	2020–2022
7-9	Tensile strength	2014–2016
7-9	Crack extension	2018–2019
7-9	Ageing	2020–2022
7-9	Size, length and angle dimensional tolerances	2014–2016
5-6	Compressive properties	2018–2020
7-9	Hardness	2018–2020
7-9	Appearance	2021–2022

Figure 2.4 *Roadmap for AM Standardization [33].*

5 PRIORITY AREAS ON ADDITIVE MANUFACTURING STANDARDS

SASAM drafted a few lists of high priority areas for AM standards after the evaluation of documents prepared by the AM community. From the common roadmap structure developed by ASTM F42 and ISO/TC261, standardization of on-going works is categorized into five priority areas as follows [33]:

- Standards for integration.
- Standards for environmental sustainability.
- Standards for quality and performance.
- Service standards.
- "Derisking" standards.

Two lists of high priority items that require attention were drafted from the feedback consolidated by SASAM. The first list consists of accessing existing standards that have relevance to the AM field and subsequently adopting it for use in AM. The second list consists of areas of interests for development determined through a survey with 122 respondents during the SASAM workshop.

The SASAM project identified a list of existing ISO standards for adoption and further development in the area of AM, and they are [33]:

- ISO/TC 61 "plastics."
- ISO/TC 106 "dentistry."
  - TC 106/SC 1 filling and restorative materials.
  - TC 106/SC 2 prosthodontic materials.
  - TC 106/SC 3 terminology.
  - TC 106/SC 4 dental instruments.
  - TC 106/SC 6 dental equipment.
  - TC 106/SC 7 oral care products.
  - TC 106/SC 8 dental implants.
  - TC 106/SC 9 dental CAD/CAM systems.
- ISO/TC 119 "powder metallurgy."
- ISO/TC 172/SC9 "electro-optical systems."
- ISO/TC 184/SC4 "industrial data."
- CEN/TC 121 "welding and allied process" (AM is partly included in committee scope).
- CEN/TC 138 "nondestructive testing."

The second list of topics and priorities are information gathered from the survey [33]. The priorities are categorized into three main topics:

product quality, materials (metal, polymer, and ceramics), and other subjects. From the list, the topics with higher priorities include:

- Product quality standards:
  - Size, length and angle dimensions, dimension tolerances standards.
  - Geometrical tolerance.
  - Tensile strength.
  - Impact strength.
  - Flexural Strength.
  - Fatigue testing.
- Material standards:
  - Co-Cr (dental, orthopedic).
  - TA6V (aeronautic repair).
  - Ti Grade 1 (medical).
  - Ti Al (aeronautic).
  - Inconel 625 (aeronautic).
  - Inconel 718 (aeronautic).
    - PA12 (medical, automotive, aeronautic, military).
    - PA flame retardant (aeronautic).
    - PA 11 (SLS).
    - ABS (FDM).
  - Other standards:
    - Post processing.
    - Monitoring process.
    - Lattice structures cleaning part recommendation/standards.

The list of priority areas, topics, and standards for AM adoption will be investigated by the AM community and standards bodies for standards development.

6 LIST OF STANDARDS AND SUMMARY

- ASTM AM standards:
  - ASTM F2924-14 standard specification for additive manufacturing titanium-6 aluminum-4 vanadium with powder bed fusion.
  - ASTM F3001-14 standard specification for additive manufacturing titanium-6 aluminum-4 vanadium ELI (Extra Low Interstitial) with powder bed fusion.
  - ASTM F3049-14 standard guide for characterizing properties of metal powders used for additive manufacturing process.

 - ASTM F3055-14a standard specification for additive manufacturing nickel alloy (UNS N07718) with powder bed fusion.
 - ASTM F3056-14e1 standard specification for additive manufacturing nickel alloy (UNS N06625) with powder bed fusion.
 - ASTM F3091/F3091M-14 standard specification for powder bed fusion of plastic materials.
 - ASTM F3184-16 standard specification for additive manufacturing stainless steel alloy (UNS S31603) with powder bed fusion.
 - ASTM F3187-16 standard guide for directed energy deposition of metals.
 - ASTM F2971-13 standard practice for reporting data for test specimens prepared by additive manufacturing.
 - ASTM F3122-14 standard guide for evaluating mechanical properties of metal materials made by additive manufacturing process.
- ISO AM standards:
 - ISO 17296-2:2015 general principles—part 2: overview of process categories and feedstock.
 - ISO 17296-3:2014 general principles—part 3: main characteristics and corresponding test methods.
 - ISO 17296-4:2014 general principles—part 4: overview of data processing.
- ISO/ASTM AM standards:
 - ISO/ASTM 52900-15 standard terminology for additive manufacturing—general principles—terminology.
 - ISO/ASTM 52915-16 standard specification for additive manufacturing file format (AMF) version 1.2.
 - ISO/ASTM 52921-13 standard terminology for additive manufacturing—coordinate systems and test methodologies.

6.1 ASTM F2924-14 Standard Specification for Additive Manufacturing Titanium-6 Aluminum-4 Vanadium With Powder Bed Fusion

This specification covers Ti-6Al-4V AM components manufactured by powder bed fusion. The components require mechanical properties similar to machined forging and wrought productions. The specification also covers related standards for powder classification, testing methods, terminology, and so forth that are applicable for the raw materials, and testing of the end product for all the required properties. To achieve final dimensions and surface finishing, post processing via machining, polishing, grinding, and so forth will be required [34].

6.2 ASTM F3001-14 Standard Specification for Additive Manufacturing Titanium-6 Aluminum-4 Vanadium ELI (Extra Low Interstitial) With Powder Bed Fusion

This specification covers Ti-6Al-4V ELI AM components manufactured by powder bed fusion. The components require mechanical properties similar to machined forging and wrought productions. The specification also covers related standards for powder classification, testing methods, terminology, and so forth that are applicable for the raw materials, and testing of the end product for all the required properties. To achieve final dimensions and surface finishing, post processing via machining, polishing, grinding, and so forth will be required [35].

6.3 ASTM F3049-14 Standard Guide for Characterizing Properties of Metal Powders Used for Additive Manufacturing Process

This guide provides the user techniques for characterizing metal powder used in AM processes. The guide references to other standards to determine test methods, practices, guides so forth for AM metallic powders. AM feedstock powders are used for a wide variety of AM processes (powder jetting, SLS, EBM, SLM, etc.). The properties of these powders need to be known to achieve products with consistent reliability and repeatability. The guide serves as a reference for stakeholders that produce, use, or sell metal powders for AM processes. This guide is also applicable to a certain extent, for powders of polymer or ceramic composition [36].

6.4 ASTM F3055-14a Standard Specification for Additive Manufacturing Nickel Alloy (UNS N07718) With Powder Bed Fusion

This specification covers UNS N07718 AM components manufactured by powder bed fusion. The components require mechanical properties similar to machined forging and wrought productions. The specification also covers related standards for powder classification, testing methods, terminology, and so forth that are applicable for the raw materials, and testing of the end product for all the required properties. To achieve final dimensions and surface finishing, post processing via machining, polishing, grinding, and so forth will be required [37].

6.5 ASTM F3056-14e1 Standard Specification for Additive Manufacturing Nickel Alloy (UNS N06625) With Powder Bed Fusion

This specification covers UNS N06625 AM components manufactured by powder bed fusion. The components require mechanical properties similar to machined forging and wrought productions. The specification also covers related standards for powder classification, testing methods, terminology, and so forth that are applicable for the raw materials, and testing of the end product for all the required properties. To achieve final dimensions and surface finishing, post processing via machining, polishing, grinding, and so forth will be required [38].

6.6 ASTM F3091/F3091M 14 Standard Specification for Powder Bed Fusion of Plastic Materials

This specification covers the requirements and component integrity of any manufactured plastic components by powder bed fusion processes, including unfilled formulations and formulations containing fillers, functional additives like flame retardant, and reinforcements or combinations thereof. It does not cover processes (SLA, FDM, LOM, etc.) that do not require the use of powder. The processes of powder bed fusion can be made reference to ASTM F2792. The specification also includes the use of additives, fillers, and reinforcement in the plastic powder [35].

For traceability purposes, plastic components fabricated by AM are classified into 3 different categories: Class I, II and III.

Class I components have the highest requirements among all the classes. The parts produced as Class I have the highest quality components that are traceable through documents produced. In testing, Class I components are subjected to the certification.

Class II components require less traceability as compared to Class I parts. The parts produced as Class II are high-quality components that do not require very detailed traceability, unlike Class I components. Class II components are subjected to the certification.

Finally, class III components are used as guidelines and require minimum traceability. Unless otherwise specified, no testing specimens are required. Class III components are usually for general usage and for early stage rapid prototyping uses. The specification also describes the fabrication of test specimen for testing purposes for Classes I, II, and III (if required).

6.7 ASTM F3184-16 Standard Specification for Additive Manufacturing Stainless Steel Alloy (UNS S31603) With Powder Bed Fusion

The manufacturing of UNS S31603 parts through powder bed fusion process based on full melt powder is covered in this standard. This standard will state the requirement of mechanical properties of the product has to be similar to machined forging and wrought products. The product also has to meet the desired surface finish and critical dimensions through post processing [39].

6.8 ASTM F3187-16 Standard Guide for Directed Energy Deposition of Metals

This guide assists the users in optimally utilizing directed energy deposition (DED) techniques for AM. It covers technology application space, process limitations, machine operations, process documentation, best work practices, etc. It defines DED as an AM process that uses energy to fuse material while they are being deposited onto a surface. There are many DED systems, including laser beam, electron beam, or plasma energy. The feedstock of DED process can be either wire or powder, which are deposited under an inert gas condition [39].

6.9 ASTM F2971-13 Standard Practice for Reporting Data for Test Specimens Prepared by Additive Manufacturing

To ensure a common database, a common format for data reporting for results by testing or evaluation of AM specimen is required. The practice describes a standard procedure for presenting data with two purposes. The first purpose is to establish further data reporting requirements, and the second is to provide a design of material property database with the essential information [40].

The practice requires users to:

- Know the minimum data element required for reporting.
- Standardize test specimen, descriptions and reports.
- Assist designers with a general standard AM database.
- Improve traceability of AM materials.
- Enable modeling and computational simulation through property-parameter-performance data captured from AM specimen.

6.10 ASTM F3122-14 Standard Guide for Evaluating Mechanical Properties of Metal Materials Made by Additive Manufacturing Process

This guide references to existing standards, if applicable, for the testing of AM metal components.

Factors that will affect the properties of AM metal components are properties of material, anisotropy, preparation of material, porosity, specimen preparation, testing environment, alignment, and gripping of the specimen, testing speed, and temperature. This standard does not cover any safety related aspects of AM, and the user shall be responsible for establishing the safety and health practices, in accordance with regulatory requirements, before prior usage [38].

The guide consists of a collection of test methods used by ASTM for the testing of material in the areas of deformation and fatigue testing.

Under deformation properties there are:

- Tension.
- Compression.
- Bearing.
- Bending.
- Modulus.
- Hardness.

Under fatigue properties there are:

- Fatigue.
- Fracture toughness.
- Crack growth.

The reporting guidelines are to adhere to standards that are applicable to each testing procedure. Due to the nature of fabrication of metallic powders in AM, anisotropic properties will be adherent to the specimen being tested. These data recorded are to be reported in accordance with ISO/ASTM 52921.

6.11 ISO/ASTM 52900-15 Standard Terminology for Additive Manufacturing—General Principles—Terminology

This terminology establishes and defines terms that are used in AM. There are currently seven different classifications of AM processes, which are as of follows [41]:

- VAT photopolymerisation.
- Material jetting.
- Binder jetting.
- Material extrusion.
- Powder bed fusion.
- Sheet lamination.
- Directed energy deposition.

6.12 ISO/ASTM 52915-16 Standard Specification for Additive Manufacturing File Format (AMF) Version 1.2

This specification describes a framework for an interchange format that will address the current and future needs of AM. STL, which is the defacto standard format, only contains mesh information and have no other provisions to provide other useful data like color, texture, material, and other properties. As AM evolves, STL will not be able to support the information database and thus this standard specification is to set the framework to establish a new format to replace STL and address the growing needs of AM to support newer features.

The new file format will be in adherence to an extensible markup language (XML) and must be able to support standards-compliant interoperability. The file format will have to describe an object in a general way that all machines can fabricate parts best to its ability.

The AMF file format must also be simple to implement and debug when required, scales well with complexity and size, have a reasonable duration for read and write operation of the file.

6.13 ISO/ASTM 52921-13 Standard Terminology for Additive Manufacturing—Coordinate Systems and Test Methodologies

This terminology describes the necessary terms used for measurement of additively manufactured specimens and also the references on the build platform [14]. This standard is intended to be, where possible, compliant to ISO 841 and assist in clarifying specific principles adapted for AM. This standard does not cover non-Cartesian systems. The terminology also cites examples of build volume origin, part rotation in the *xyz* coordinates and minimum perimeter bonding box, for the user's reference.

6.14 ISO 17296-2:2015 General Principles—Part 2: Overview of Process Categories and Feedstock

ISO 17296 Part 2 describes the general process and working principles of AM machines [42]. There are many different terminologies used in AM processes, and it creates confusion when a user wants to know the working principle behind a particular machine. This ISO standard classifies part quality based on certification, testing and traceability into three classes, while process chains are classified into two categories, namely, single- and multistep process. The processes are further divided into seven different categories which are:

- Vat photopolymerization.
- Material jetting.

- Binder jetting.
- Powder bed fusion.
- Material extrusion.
- Direct energy deposition.
- Sheet lamination.

6.15 ISO 17296-3:2014 General Principles—Part 3: Main Characteristics and Corresponding Test Methods

ISO 17296 Part 3 documents the testing requirements of components that are fabricated by AM. It covers the quality characteristics of the fabricated components, test procedures, scopes, content of test and supply agreements [43].

Characteristics of components are developed into two main parts: feed stock and part requirements. The requirements of the feedstock of the bulk material are:

- Powder particle size.
- Morphology.
- Surface and distribution.
- Tap and apparent density.
- Flowability.
- Ash content.
- Carbon content.

For the part requirements, they are separated into four main components: surface, geometric, mechanical, and build material requirements. Specifically:

- Surface requirements: appearance, surface texture, and color.
- Geometric requirements: size, length, angle, tolerances, geometrical tolerancing.
- Mechanical requirements: tensile, impact, compressive, flexural, and fatigue strength, hardness, creep, gaining, friction coefficient, shear resistance, and crack extension.
- Build material requirements: density, physical, and physico-chemical properties.

Testing of AM components is divided into three main categories, which are safety critical, nonsafety critical and prototyped parts. The ISO document contains three tables for the testing requirements, namely, one that lists the tests that have to be fulfilled, one for tests that are recommended, and lastly one for tests that are not applicable. The tests would be subject to agreement between the part vendor and the customer.

The standard covers a full range of related standards for the requirements of bulk material, testing procedures and requirements of the end product.

6.16 ISO 17296-4:2014 General Principles—Part 4: Overview of Data Processing

ISO 17296 Part 4 is targeted at users of AM systems and related software systems. It describes existing data formats used in AM, and targets at users from the computer aided design/computer aided engineering (CAD/CAE) perspective to reverse engineering companies, testing bodies, and production of additive manufacturing systems and software [43].

Found in this ISO standard are the terminologies used in the construction of the 3D model and the process from which the 3D model is polygonised and sliced into layers. Some data formats used in this ISO are STL, VRML, IGES, VDA-FS, STEP, and AMF. More information on these data formats can be extracted from the ISO document.

This ISO also sets the requirements for data quality. Data quality is important as they will determine if the 3D printed part is of high quality. To achieve high quality object, surfaces of models must blend smoothly and trim to achieve a watertight model, and be orientated for easy volume identification by the software. During triangulation, no construction aids shall be selected and all surface models must be converted into solid volumes before triangulation/polygonization (creation of polygons in an encased volume). Included inside the ISO standard also documents the potential formatting errors in the STL data.

7 QUESTIONS

1. Which organizations are developing the standards for AM?
2. Why conventional manufacturing standards are not suitable for AM applications?
3. Discuss about the importance of standardization in the AM industry.
4. Who is the first official standard body for AM?
5. What is the common structure of AM standards?
6. What are the goals in SASAM roadmap for AM standardization?

REFERENCES

[1] B. Kraemer, K. Bartleson, J. Handal, "Importance of standards for industry practitioners," in: IEEE Sections Congress, Amsterdam, 2014.

[2] IEEE Standards Asociation. (2011, 30/12/2016). What are standards? Why are they important? Available from: http://standardsinsight.com/ieee_company_detail/what-are-standards-why-are-they-important
[3] M.D. Monzón, Z. Ortega, A. Martínez, F. Ortega, Standardization in additive manufacturing: activities carried out by international organizations and projects, Int. J. Adv. Manufactur. Technol. 76 (2015) 1111–1121.
[4] J.-Y. Lee, W.S. Tan, J. An, C.K. Chua, C.Y. Tang, A.G. Fane, T.H. Chong, The potential to enhance membrane module design with 3D printing technology, J. Memb. Sci. 499 (2016) 480–490.
[5] B. Dutta, F.H. Froes, 24—the additive manufacturing (AM) of titanium alloys, in: M.Q.H. Froes (Ed.), Titanium Powder Metallurgy, Butterworth-Heinemann, Boston, 2015, pp. 447–468.
[6] D. Shi and I. Gibson, Surface finishing of selective laser sintering parts with robot, *Solid Freeform Fabric.*, In Proceedings of the 9th Solid Freeform Fabrication Symposium, Austin, Texas, 1997, pp. 27–35.
[7] H. Gong, K. Rafi, H. Gu, T. Starr, B. Stucker, Analysis of defect generation in Ti–6Al–4V parts made using powder bed fusion additive manufacturing processes, Add. Manufactur. 1–4 (2014) 87–98.
[8] 3D Printing Industry. (2016). History of 3D printing. Available from: http://3dprintingindustry.com/3d-printing-basics-free-beginners-guide/history
[9] T. Caffrey, T. Wohlers, Wohlers report 2016, Wohlers Associates, Inc, Colorado, USA, (2016).
[10] S. Tranchard V. Rojas. (2015). Manufacturing our 3D future. Available from: http://www.iso.org/iso/news.htm?refid=Ref1956
[11] 3D Systems Inc. (2016), What is an STL file? Available from: https://www.3dsystems.com/quickparts/learning-center/what-is-stl-file
[12] J. Munguía, J.d. Ciurana, C. Riba, Pursuing successful rapid manufacturing: a users' best-practices approach, Rapid Prototyp. J. 14 (2008) 173–179.
[13] J. D. Hiller H. Lipson, STL 2.0 A proposal for a universal multi-material additive manufacturing file format, in: Proceedings of the Solid Freeform Fabrication Symposium, In Mechanical and Aerospace Engineering, ed, 2009, pp. 266–278.
[14] ISO and ASTM, Standard terminology for additive manufacturing—coordinate systems and test methodologies, in ISO / ASTM52921-13, Standard Terminology for Additive Manufacturing-Coordinate Systems and Test Methodologies, ASTM International, West Conshohocken, PA, 2013.
[15] Stratasys. (2015). ASTM additive manufacturing standards what you need to know, Available from: https://www.stratasysdirect.com/blog/astmstandards/
[16] K.K. Jurrens, Standards for the rapid prototyping industry, Rapid Prototyp. J. 5 (1999) 169–178.
[17] ASTM. (2015). What is a work item? Available from: http://www.astm.org/DATABASE.CART/whatisaworkitem.html
[18] ISO. (2015). Technical committees—ISO/TC 261—Additive manufacturing, Available from: http://www.iso.org/iso/iso_technical_committee?commid=629086
[19] CEN/TC 438, Business plan CEN/TC 438 additive manufacturing executive summary, European Committee for Standardization (CEN), 2015.
[20] ISO/TC 261 and ASTM F42, Joint plan for additive manufacturing standards development, ISO and ASTM International, 2013.
[21] ASTM. (2015). Additive manufacturing technology standards. Available from: http://www.astm.org/Standards/additive-manufacturing-technology-standards.html
[22] E. Pei, ISO TC 216 WG4 presentation to SMF by Dr Eujin Pei, presented at the ISO TC 261 WG 4, Singapore, 2015.
[23] ISO. (2015). ISO Standards - ISO/TC 261 - Additive manufacturing. Available from: http://www.iso.org/iso/home/store/catalogue_tc/catalogue_tc_browse.htm?commid=629086&published=on&includesc=true

[24] ISO and CEN, Agreement on technical co-operation between ISO and CEN (Vienna Agreement), ASTM International, 1991.
[25] D. L. Bourell. (2013). NIST roadmapping workshop: Roadmaps for additive manufacturing—past, present, future. Available from: http://events.energetics.com/nist-additivemfgworkshop/pdfs/Plenary_Bourell.pdf
[26] New agreement strengthens partnership between ISO and ASTM on additive manufacturing, ISO, 2011.
[27] ANSI ISO and ASTM to cooperate on international standards for additive manufacturing, New York, USA: ANSI, 2011.
[28] ASTM and ISO additive manufacturing committees approve joint standards under partner standards developing organization agreement, ISO, ASTM, 2013.
[29] P. Picariello, Presentation on collaboration for AM standards development, ASTM International (2015).
[30] Additive manufacturing community to meet in Nottingham (2014). Metal Powder Report, pp. 33–36.
[31] SASAM, Final report summary—SASAM (Support Action for Standardisation in Additive Manufacturing), SASAM, 319167, 2014.
[32] F. Feenstra, K. Boivie, B. Verquin, A. Spierings, H. Buining, M. Schaefer, et al., SASAM D2.3 Final version of roadmap for AM standardisation, SASAM, 2013.
[33] SASAM, Additive manufacturing: SASAM standardisation roadmap 2015, SASAM, 2015.
[34] ASTM, Standard specification for powder bed fusion of plastic materials, in: F3091(M)-14, ed: ASTM International, 2014.
[35] ASTM, Standard specification for additive manufacturing Titanium-6 Aluminum-4 Vanadium ELI (Extra Low Interstitial) with powder bed fusion, in: F3001−14 ASTM International, 2014.
[36] ASTM, Standard guide for characterizing properties of metal powders used for additive manufacturing processes, in: ASTM F3049-14, ASTM International, 2014.
[37] ASTM, Standard specification for additive manufacturing nickel alloy (UNS N07718) with powder bed fusion, in: ASTM F3055-14a, ASTM International, 2014.
[38] ASTM, Standard specification for additive manufaturing nickel alloy (UNS N06625) with powder bed fusion, in: ASTM F3056-14 ASTM International, 2014.
[39] ASTM, Standard specification for additive manufacturing stainless steel alloy (UNS S31603) with powder bed fusion, in: ASTM F3184-16, ASTM International, 2016.
[40] ASTM, Standard practice for reporting data for test specimens prepared by additive manufacturing, in ASTM F2971-13, ASTM International 2013.
[41] ISO and ASTM, ISO/ASTM 52900:2015 Additive manufacturing—General principles—Terminology, ISO, ASTM International, 2015.
[42] ISO, Additive Manufacturing General Priniciples Part 2 Overview of process categories and feedstock, in: 17296-2-2015, ISO, 2015.
[43] ISO, Additive Manufacturing General Principles Part 3 Main characteristics and corresponding test methods, in: 17296-3-2014, ISO, 2014.

CHAPTER THREE

Measurement Science Roadmap for Additive Manufacturing

Contents

1 INTRODUCTION TO MEASUREMENT SCIENCE IN ADDITIVE MANUFACTURING

AM is expected to have a significant impact on the US economy, due to its ability to produce high value, high quality, and complex parts, while reducing turnovers and manufacturing cost when compared to conventional machining [1]. In 2011, US industry made USD 246 million from AM shipments alone [2]. Although AM technology has significantly advanced over the past few years, its adoption has been sluggish as there are many barriers restricting companies from investing in the technology.

There are numerous challenges to overcome before companies consider investing in AM technology. The challenges range from lack of variety of material, poor part accuracy, poor repeatability and consistency, and the lack of qualification and certification standards [1]. Through workshops with the community, the National Institute of Standards and Technology (NIST) has generated action plans and road maps that address these challenges, which are categorized into four main topics [3].

The challenges are:

- Material uncertainties.

Standards, Quality Control, and Measurement Sciences in 3D Printing and Additive Manufacturing
http://dx.doi.org/10.1016/B978-0-12-813489-4.00003-9

- Process uncertainties.
- Part accuracy and uncertainties.
- Physics-based and property-based models for simulation and analysis.

One example that showcases the commitment of NIST toward advancing AM measurement science is the value of grants given out to research communities. In 2013, NIST awarded two grants totaling up to USD 7.4 million to fund projects on related subjects and standards to promote research in the area of measurement science for AM. Out of the USD 7.4 million, USD 5 million was awarded to National Additive Manufacturing Innovation Institute (NAMII, now known as American Makes) and the rest to Northern Illinois University [4].

NIST also actively engages the AM community to address any challenges faced. Issues are raised for discussion during workshops and road maps are formulated to address these gaps in measurement science [5]. The AM roadmap created by NIST in recent years, based on prior roadmaps, aims toincrease adoption of AM measurement science in the industry [3]. The goal of the roadmap is to:

- Develop standards and protocols for all aspects of AM, from material design and use, to part build and inspection.
- Develop measurement and monitoring techniques and collect data, from material feedstock through final inspection, including effective process controls and feedback.
- Characterize material properties, which are keys to material development, processing effectiveness and repeatability, qualification of parts, and modeling at many levels.
- Create modeling systems that couple design and manufacturing, which impacts the development of materials, as well as new processing technologies.
- Introduce closed-loop control systems for AM that are able to monitor and correct the process in real time, which are vital for processing and equipment performance, assurance of part, adherence to specifications, and the ability to qualify, and certify parts and processes.

Publications developed by NIST address some of the measurement needs in the AM industry. By developing test protocols, procedures and analysis methods via round-robin testing of AM materials [2], NIST hopes to improve the confidence level of industry users to adopt AM.

Currently, AM technologies are used mainly for rapid prototyping. To introduce AM as a manufacturing process, the risk of using AM must be mitigated to a level acceptable to stakeholders in any company. As AM measurement science and standards are still in its infancy, NIST needs to build its capabilities for measurement science.

2 CHALLENGES OF MEASUREMENT SCIENCE

AM components are not widely used in the industry mainly due to lack of certification and qualification methods, which arises from the lack of established measurement science and standards. This is a result of uncertainties in both material and processes, as shown in Fig. 3.1 [6].

There is limited knowledge pertaining to the specific properties that define the performance of the material used in a particular AM system or process. Unlike industrial AM machines which are usually developed as a closed platform, the hardware and software of consumer grade AM systems are typically open-source and can be easily modified. Manufacturers of industrial AM systems primarily rely on empirical methods during the development process, such as for hardware design and process optimization, largely due to the lack of published work related to the development of AM systems. Many companies that have developed industrial grade systems are unwilling to share their technologies and patents freely due to the countless hours of research and funding devoted to achieve the level of precision needed for the manufacturing industry.

Owing to a large amount of capital and time invested in such a development process and a limited market, early adopters who pioneered AM development (3D Systems, Stratasys, etc.) have kept the technology proprietary. The hardware and process parameters of such industrial grade systems are therefore hard to modify. Furthermore, it is often not possible to use third party products like powder or resin [7]. This has resulted difficulties in implementing AM process monitoring systems by retrofitting existing industrial AM systems with sensors and measurement devices to monitor the AM process.

The development of measurement science is, therefore, difficult due to these main factors:

- AM technologies, being relatively new and emerging, require collaboration between different disciplines of studies. Integration of different fields is important for the optimization of AM machines. Material knowledge for AM systems is limited, making it difficult to produce high-quality parts.

Figure 3.1 ***Uncertainties in raw material and system will result in more uncertainties in final part [6].***

- AM systems are complex in nature, where optimized AM systems have customized software and hardware that work in tangent with one another. Evaluation of these machines is, therefore challenging.
- Many AM machines are built as "black boxes" with technological hardware hidden from any end users and process developers. In order to integrate new software and hardware, process developers will have to work with AM vendors right from the early machine design and development stage to ensure proper optimization of their system.

AM machines must be "open-source" to allow better integration of third party sensors, equipment and software to monitor process parameters and correct the manufacturing process in real time. With proper sensors and measurement devices for monitoring and feedback, there would be better assurance of the consistency of the fabricated part and the AM process.

2.1 Additive Manufacturing Materials and Uncertainties

The variety of AM materials used in today AM systems are limited, with the majority of material being polymer or metal based [8]. Due to the lack of research on both material properties and material manufacturing, developing new materials for AM is time-consuming as it requires empirical experiments. The common understanding of typical raw materials in their bulk form cannot be applied onto their AM equivalents in powder form because at micron level, tertiary forces such as interparticle friction plays a more signification role compared to in the bulk material. Therefore, a greater understanding of both AM raw materials and processes is needed to predictably estimate the properties of an AM fabricated part. There are several additional factors unique to the AM process to consider in order to predict the final material properties, such as the microstructure and powder morphology, resin curing time and light sensitivity for resin-based materials, filament consistency, composition, and melting point for polymeric FDM material.

The current technologies adopted in powder based AM systems to capture powder characteristics are not adequate. For example, the use of hall-effect flowmeters to capture particle size assumes that the particles are spherical. This may not accurately represent the size if particles are oddly shaped [9]. The accuracy of laser-based measurement systems to characterize powder is also limited due to the variation in the reflective and refractive properties among different materials. The current techniques employed for powder property measurement thus tend to be inadequate for characterization.

Microstructure of raw material used in AM is crucial in determining the properties of the final part. Most components made by powder bed fusion (PBF) systems retain the original microstructure of the raw material, without any postprocess treatment. Experiments conducted by two different research groups pointed out that the microstructure of raw stainless steel powders did not differ from that of the final component produced by an AM system in both argon and nitrogen inert gas environments [9]. Instead, it was found that the powder microstructure is determined solely during atomization [9]. Furthermore, no two powder manufacturers will be able to produce identical microstructure if they use different processes. Knowledge of the microstructure of the powder will aid in predicting the effect on the strength and quality of the final part. Hence, the qualification of AM components produced will be difficult unless there are proper characterization and certification of the raw material. Any lack of information of the raw material creates difficulty in assessing the properties of that component to ensure that it is within a customer's specified requirements.

Studies have shown exceptions where microstructures formed during AM process may change, owing to different powder compositions [9]. However, the control of these microstructures is difficult. Rapid cooling rates during the sintering process may also result in poor bonding ability between layers of powder, leading to poor densification of the component [10]. This poses issues to the component, as it may shrink below the allowable tolerance and also cause embrittlement.

Without proper documentation of materials used in AM, parts produced have to be subjected to destructive testing to ensure its worthiness. Unfortunately, the AM database for material does not have enough material information for adequate and proper crossreferencing, resulting in expensive and time-consuming iterative works to determine the correct parameter for any AM process.

A database of material would establish baselines, along with certification methods to classify powders, through new techniques that measure size distribution, mechanical properties, and microstructures etc., specifically tailored towards AM. Knowledge is also lacking in the area of surface roughness of parts produced by AM. It would be possible for surface roughness to be characterized by identifying the key factors affecting it, such as the specific process parameters or material properties. However, no such library presently exists for reference that correlates surface roughness to process/material type.

Postprocessing of components produced by AM is usually required to achieve the desired dimensions, strength and other required properties. Some postprocess treatments for densification of parts or reducing residual stresses are hot isostatic pressing (HIP) and heat treatment. Although they are well documented for treating parts fabricated by conventional methods, these postprocess treatments are not well understood in the context of AM [3].

2.2 Additive Manufacturing Processes and Uncertainties

Most AM systems developed in this decade lack complex tools and sensors to measure process performance. Current technologies employed by AM vendors are inadequate for any forms of in situ measurement, monitoring and control of part production. Unless process parameters for a particular set of powders on a specific system are provided by AM vendors, printing would be based on empirical studies.

The feedback systems in most AM machines are considered "open-loop". In these machines, the print process will stop only through human intervention when a fault is detected manually, and only through human intervention will the problem be resolved. In other words, if the fault is not detected by the user, the machine will continue printing until the component is beyond recovery [11]. It is, therefore, important to monitor the process and control accordingly—there is currently little or no benchmarking of the level of monitoring and control employed for AM processes.

Measurement and monitoring of the materials during fabrication is critical to detecting defects (such as voids, inclusions, and high thermal gradient) in the AM components. The defects can be detected and rectified in situ, which will save time and cost of reproducing a defect-free component. However, defect detection is only possible with in situ measurement and monitoring capability, which unfortunately most systems lack of.

Most commercial AM machines are built as closed systems which restrict installation of additional sensors, high-speed cameras, and thermographs on their systems by end users without voiding warranty or manufacturer's support. Furthermore, sensors are not cheap and the integration of these sensors will drive up the production cost of the AM systems. Software integration is also a major challenge especially for closed systems unless explicitly supported by the machine manufacturer. The algorithms in present-day AM software are also relatively simple. They generally do not have in situ corrective process management feature which provides automatic compensation abilities to reduce the possibility of defects in the component.

The lack of understanding and ability to control stress formation during the process also affects the end product. Without any monitoring, stress residue could have unknowingly formed at certain areas of an additively manufactured component that may result in early failure. The ability to monitor process anomalies is therefore crucial to assuring the final properties of the parts produced by AM.

Sensors, measurement devices and algorithms are required to measure and predict the following properties of AM components during fabrication:

- Dimensions.
- Geometry.
- Roughness.
- Surface finish.
- Structural (microstructure and mesostructure).
- Defects (porosities, flaws, warpages, and etc.).
- Energy source measurement (amount of energy to create a melt pool).
- Temperature range.

Any sensors or measurement devices have to be calibrated to within an acceptable margin of error or uncertainty as they are critical to ensuring the quality of parts produced by AM. To control part uncertainties, measurement is required throughout the entire process, from the raw material, to during fabrication, and lastly when the product is completed. This provides a comprehensive insight to the performance of the AM system and process [12].

Therefore, it is important to develop new sensors, models and measurement methods, which are crucial for closed-loop systems. It is essential to accurately sense, measure, and actively control the process real-time with a robust algorithm to achieve part uniformity and consistency. This measurement and characterization has to be extended to include raw material properties, as well as postprocessing methods to achieve a more accurate diagnosis. The integration of process control and feedback into AM systems will inevitably produce higher quality parts with greater assurance of performance properties.

2.3 Additive Manufacturing Parts and Uncertainties

Quality inspection tools are employed to measure critical aspects such as tolerances of a dimension and the mechanical properties etc. of additively manufactured parts. As the inspection tools are generally developed for conventionally manufactured parts, they may not be adequate to inspect all aspects of additively manufactured parts, such as measurement of internal porosities or complex internal structures. Costly inspection tools, such as ultrasound and X-ray may be needed for such cases.

For PBF systems, which usually employ a laser or electron beam energy source, beam quality is one of the key factors determining the quality of the final component. With the installation of appropriate sensors, the quality of the laser or electron beam can be monitored and controlled during fabrication, ensuring better quality components.

From the measurement of beam quality, assessment can be made on AM systems to determine their performances and differences between similar machine models. For example, both machines may employ laser sintering but comprise different subcomponents and build configurations. Furthermore, machines used to fabricate similar parts may not necessarily exhibit similar tolerances when it comes down to the accuracy of the part. Every machine possesses its own characteristics and behavior that will result in part variations not detectable by the naked eye. Systems from the same manufacturer or vendor but assembled in different locations may also result in inconsistencies in the built parts.

2.4 Additive Manufacturing Standards

Standards are driven by the needs of the industry and their development is usually slow due to high demand in both budget and time, and the lack of volunteers. Today, standards related to AM are far and few, with several on-going works—the majority of which are focused mainly on metallic powder [13].

There are limited standards for raw materials used in AM, as well as the processes involved in their fabrication. Current standards for metallic powder are predominantly on steel based alloy (UNC) and Ti-6 Al-4 V, which are processed using PBF. There are no standards on other type of metallic materials.

Standards for the polymer-based material are very general, and there are no open standards developed for resin- and filament-based AM. In the case of polymer powders, a standard on characterization is made for PBF powders [14]. This standard only applies to powder-based polymers and not wire-fed polymers used in FDM. It is clear that there are gaps in standards defined for raw materials of AM. AM standards related to material processing only cover a select few materials and processes, leaving no guidelines for the vast majority of materials and processes available.

Developing standards for materials used in AM is challenging, as the material formulated by a particular vendor may either be patented or its formula kept secret due to commercial interests. Currently, only one official standard that looks at metal powder characterization for AM [15]. However, there is no official powder characterization standard for polymers and ceramics. Although polymer powder suppliers usually provide the material properties, such as strength and flexibility, along with the packaging, a side by side comparison of materials from different sources is usually difficult due to the lack of a defined set of standard test parameters. This also hinders the possibility of a common database where one can refer and compare the powder properties and check for any compatible materials. Without overcoming these challenges, it is difficult to optimize AM systems to produce high-quality parts [7].

A list of standards that are lacking for AM is summarized below:

- Measurement and characterization of AM processes.
- Measurement test pieces.
- Machine consistency.
- Calibration standards.
- Characterization of multimaterial parts.
- Data reporting and collection.
- Process parameters.

Standards are developed faster in areas with greater industry interest compared to areas of lesser interest where there is little or no research done. As the AM industry matures, standards development will gain traction over time and more standards are expected to be published to address the changing needs of the community.

2.5 Additive Manufacturing Modeling and Simulation

To optimize the cost of manufacturing AM component parts, modeling and simulation of the component and the AM process by means of physics-based simulation will help to determine the feasibility of producing the end component without empirical trials. Potential issues that may arise during the AM process, such as excessive deformation, unsupported overhanging structures, etc. can be predicted by simulations and therefore to be addressed prior to actual fabrication. This will be only feasible if there are databases and predictive models that can be used for AM simulation before the parameters are imported over to the AM system.

With physics-based simulations, it is possible to predict the properties (microstructures, defects, surface topology, residual stress etc.) of the final component, without physically fabricating it. For example, part distortion, which is caused by poor temperature control or how the part is oriented, can be eliminated with a proper understanding of the process through simulation.

Although simulation software will cut time and cost by reducing the number of physical prints needed for empirical trials, it is only as good as the algorithms and models of AM elements input into the system. Until there are proper models developed for the simulation software, the AM community would have to resort to trial prints to determine fabrication feasibility.

3 POTENTIALS OF MEASUREMENT SCIENCE

Measurement science certifies the worthiness of components to be used in commercial and industrial products, and in turn promoting competitiveness between companies. To overcome its challenges, standards and techniques have to be established for adoption. These challenges are taken on by federal agencies such as NIST that develops new test methods, processes, and artifacts for AM. By establishing common benchmarking guidelines, improvements in AM measurement science would foster greater competition between companies as customers will be able to fairly compare between different AM systems. These goals are achievable through areas, such as material characterization, real-time process control, process and product qualification, and systems integration.

A materials database for AM will consist of a large library of material properties, knowledge of material, process parameters and possible defects [2]. With a proper materials database, side by side comparison of materials would be convenient. Processing parameters can be easily ported from one machine to another, as long as the machines are capable of processing that particular type of raw material.

Materials for AM can be divided into 3 main types based on form: powder-, liquid-, and solid-based. They may also be further classified into metals, polymers and ceramics. Properties such as microstructure, strength and part density, can be crossreferenced and compared. Cost could also be added so that users can select materials within their budget. A materials database could also link any raw material to a list of known vendors so as to reduce the time taken for sourcing of suppliers which in turn will generate a higher turnover for an AM company.

To achieve optimal part quality, in situ monitoring with active process measurement and control is vital. Hu conducted an experiment using a charge-coupled device (CCD) camera as a sensing device to optimize laser cladding process in an AM system [16]. In that experiment, a consistent melt pool width was achieved with closed-loop feedback monitoring of the melt pool by means of the CCD camera [16]. This resulted in more uniform microstructures and evenly distributed thermal residual stress in the component after the process [16]. The closed feedback loop ensures predictable and consistent part quality. Processes such as DMLS, SLM, and EBM will be able to achieve better part quality with a high-speed feedback control as any likely defects would be detected in real time during printing, and the algorithms will "correct" the potential defect in the next layer.

Nondestructive examination (NDE) can be employed into AM for postprocess measurements. Current NDE techniques (X-ray, ultrasound etc.) that are used in the aerospace, medical, and other industries that make unique parts unsuitable for destructive testing can be adopted for use in AM. For example, laser-based microscopy can be used to measure surface topology, while coordinate measuring machine (CMM) can measure part size and dimensions. It is important to note that current NDE techniques may not be adequate for AM and hence new NDE techniques may be required.

Modelling and simulation have been widely used in conventional manufacturing to predict the properties of the bulk material, as well as the final product. Finite element analysis (FEA), for example, is one such technique that provides engineering information, such as stress, strain, deformation, natural frequencies, etc. about a component which cannot be obtained by using traditional analysis methods. The same approach could be employed in AM to determine the effects of the process and material on the final component.

Simulating an AM process before printing will also assist in detecting the type of defects that may occur with that particular set of process parameters. Physics-based simulation of the SLM process can be used to predict the mechanical properties of a solidified melt pool. With the right physics-based algorithms, many characteristics such as residual thermal stress, surface roughness, and even microstructures can be approximated through the use of simulation programs.

Shiomi [17] performed a FEA on the melting and solidifying processes of metallic powders by laser rapid prototyping. The simulated weights of the solidified powders caused by several pulses of the laser beam agree well with the experimental data and the results provide useful information on the process parameters. This eliminates the need for multiple test print trials to determine the optimal process parameters to achieve the desired part properties.

Laser parameters could be adjusted to control the size of the melt pool, and in turn affecting the part quality. With infra-red cameras installed, the system will be able to constantly monitor and correct the laser parameters to maintain the desired geometry of the pool.

In an experiment conducted by Vegard Brotan, a laser sintering machine was simulated and corrected with an algorithm designed to correct the XY-plane accuracy and laser position [18]. The results show an overall improvement in the roundness of shafts that were sintered, especially those away from the center of the powder bed. This is one of the many experiments aimed at improving AM processes to achieve better part quality.

Advancements in measurement science for AM will help to define criteria for certification and qualification of parts. This is key to instil confidence in AM especially in critical industries. Measurement science also enables development of more consistent and reliable AM processes by means of closed-loop monitoring and control systems. It also involves simulation models and promotes a good understanding of raw material properties and process parameters resulting in a consistent and reliable printed part.

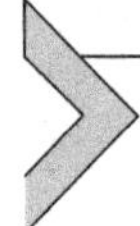

4 WORKS PUBLISHED BY NATIONAL INSTITUTE OF STANDARD AND TECHNOLOGY

NIST has published on their website several related works that they have completed, many of which are beneficial to stakeholders in the AM industry. One such development is a test artifact that can be used to determine a machine performance level. Another article defines guidelines on the testing methods focused on polymer-based prints [8].

The performance of AM machines has to be determined to ensure that parts produced across different machines are within specifications. To address this, NIST has developed a test artifact model [19,20] which has a range of features of different sizes, height, and roundness to assess the performance and capabilities of the AM machines.

Some of the features incorporated in the test artifact are [19]:

- Diamond shaped base.
- Staircases and vertical faces of staircase.
- Pins and holes.
- Fine features.
- Central hole and cylinders.
- Ramp.
- Lateral features.
- Top surface.
- Outer edge.

The features in the test artifact are used to measure a range of abilities of a particular set of AM machines. The features will test for [19]:

- Flatness and warping.
- *Z* axis linear step accuracy.
- *X* axis alignment and parallelism.
- *Y* axis alignment and parallelism.
- Roundness and concentricity.
- Straight edge.
- Fine features observable by microscope and minimum feature size.
- Overhanging features without support structures.
- 3D contours.
- Errors in beam size.

This test artifact has been developed by consolidating features from a range of previous test artifacts by NIST and is intended to be used as a determinant of the capability of machine and process parameters used in an AM process. Examples of such similar artifacts are shown in both Figs. 3.2 and 3.3.

The first artifact in Fig. 3.2 consists of the following features:

- Thin walls of varying thickness.
- Staircase features.
- Angular plates of varying angles.
- Lateral features.

The second artifact in Fig. 3.3 consists of the following features:

- Pins of various diameters.
- Cylinders of various steps and diameters.
- Holes of various diameters.

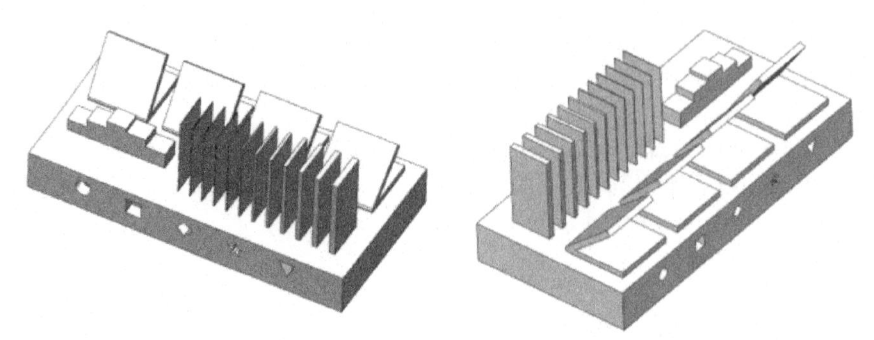

Figure 3.2 ***Artifact 1 Developed for Measurement of Angle of Print, Wall and Height Thickness.***

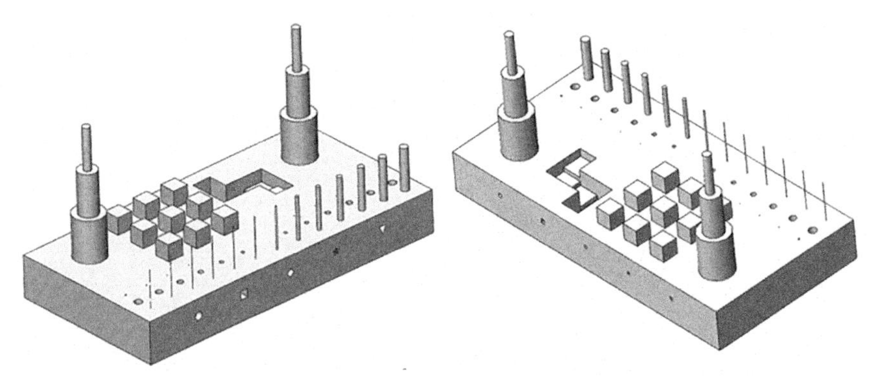

Figure 3.3 ***Artifact 2 Developed to Measure Concentricity, Step Level, Height of Pin and Single Pin Thickness.***

- Cube features.
- Staircase features.
- Lateral features.

Some of these features are similar to those available in the NIST artifact [19], while some are designed, like the angular plates, to test the limits of the printing process. The development of these artifacts will allow better understanding of the machine capability.

NIST has also published an article for the testing of polymeric materials used in AM. A survey conducted has shown results that polymers are the largest group of materials used in AM, as shown in Fig. 3.4.

The majority of the materials used in AM are polymer-based, and typically there is not much information on the performance properties mentioned in most data sheets. Comparison between datasheets of different vendors is often confusing, owing to lack of standardization. In spite of

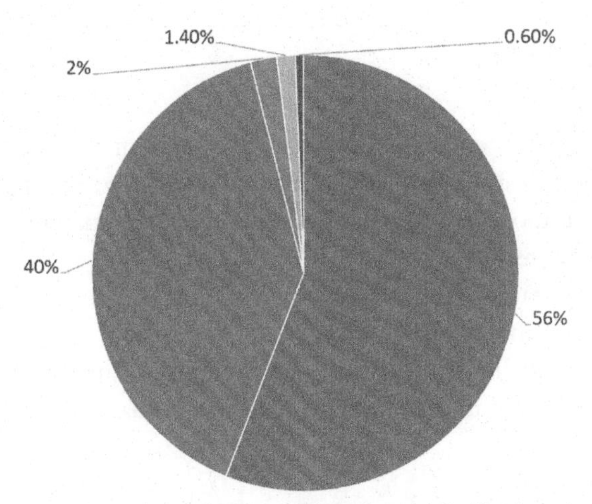

Figure 3.4 ***Percentage of Types of Materials Available in AM [8].***

citing similar references to a testing standard, tests conducted by different vendors may be conducted under different environments [8].

Regardless of the shortfalls, many current standards are widely used to determine various properties of parts produced by AM and they are:

- Tension.
- Flexure.
- Compression.
- Shear.
- Creep.
- Fatigue.
- Fracture toughness.
- Impact.
- Bearing strength and open hole compression.

The article summarizes all the test standards that can be used to determine the strength of polymer parts produced by AM, but it should be noted that there are various classifications given to the standards. Some tests prescribed in the standards are directly applicable to additively manufactured parts. However, some require postprocessing to be performed before the tests should be carried out. There are also other standards that are not applicable to AM as it would not be possible to produce the specimens prescribed in these standards with AM technologies [8]. Users of these standards need to take note of the tests that are specifically prescribed for AM specimens.

5 QUESTIONS

1. List the challenges encountered with measurement science for AM.
2. Name somemeasurements that are unique to AM components?
3. Discuss the importance of measurement science in enabling closed loop feedback systems.
4. What are the factors restricting the adoption of closed loop feedback systems in commercial AM systems?
5. What are the challenges involved in defining standards for AM?
6. What is the role of measurement science in standards development?

REFERENCES

[1] Y. Huang, M.C. Leu, J. Mazumder, A. Donmez, Additive manufacturing: current state, future potential, gaps and needs, and recommendations, J. Manuf. Sci. Eng. 137 (2014) 014001.

[2] C. Brown, J. Lubell, R. Lipman, Additive manufacturing technical workshop summary report, NIST, 2013.

[3] Energetics Inc and NIST, Measurement science roadmap for metal-based additive manufacturing, NIST, 2013.

[4] NIST, NIST awards $7.4 million in grants for additive manufacturing research, NIST, 2013.

[5] K. Jurrens, NIST measurement science for additive manufacturing presented at the PDES, Inc. technical workshop, Gaithersburg, MD, USA 2013.

[6] J.A. Slotwinski, Additive manufacturing at NIST presented at the The Science of Digital Fabrication, Massachusetts Institute of Technology, Cambridge, MA, USA, (2013).

[7] NIST, NIST Measurement science for additive manufacturing program, 2014. Available from: http://www.nist.gov/el/isd/sbm/msam.cfm

[8] A. M. Forster, Materials testing standards for additive manufacturing of polymer materials: state of the art and standards applicability, NIST, 2015.

[9] J.A. Slotwinski, E.J. Garboczi, Metrology needs for metal additive manufacturing powders, J. Minerals Metals Materials Soc. 67 (2015) 538–543.

[10] Y.-A. Jin, Y. He, J.-Z. Fu, W.-F. Gan, Z.-W. Lin, Optimization of tool-path generation for material extrusion-based additive manufacturing technolog, Addit. Manuf. 1–4 (2014) 32–47.

[11] G.J. Schiller, Additive manufacturing for aerospace, in: IEEE Aerospace Conference, MT, USA, 2015, pp. 1–8.

[12] M. Mani, B. Lane, A. Donmez, S. Feng, S. Moylan, R. Fesperman, Measurement science needs for real-time control of additive manufacturing powder bed fusion processes, NIST, 2015.

[13] Stratasys, ASTM additive manufacturing standards: What you need to know? 2015. Available from: https://www.stratasysdirect.com/blog/astmstandards/

[14] ASTM, Standard specification for powder bed fusion of plastic materials, in: ASTM F3091/F3091M-14, ASTM International, 2014.

[15] ASTM, Standard guide for characterizing properties of metal powders used for additive manufacturing processes, in: ASTM F3049-14, ASTM International, 2014.

[16] D. Hu, R. Kovacevic, Modelling and measuring the thermal behaviour of the molten pool in closed-loop controlled laser-based AM, Proc. Institut. Mech. Eng. Part B J. Eng. Manuf. 217 (2003) 441–452.
[17] M. Shiomi, A. Yoshidome, F. Abe, K. Osakada, Finite element analysis of melting and solidifying processes in laser rapid prototyping of metallic powders, Int. J. Machine Tools Manuf. 39 (1999) 237–252.
[18] V. Brøtan, A new method for determining and improving the accuracy of a powder bed additive manufacturing machine, Int. J. Adv. Manuf. Technol. 74 (2014) 1187–1195.
[19] S. Moylan, J. Slotwinski, A. Cooke, K. Jurrens, M.A. Donmez, Proposal for a standardized test artifact for additive manufacturing machines and processes, in: Solid Freeform Fabrication Symposium, Austin, TX, USA, 2012.
[20] S. Moylan, A. Donmez, D. Falvey, B. Lane, NIST qualification for additive manufacturing materials, processes and parts, 2013. Available from: http://www.nist.gov/el/isd/sbm/qammpp.cfm

CHAPTER FOUR

Software and Data Format

Contents

1 DATA FORMAT IN ADDITIVE MANUFACTURING

The fabrication of parts and components by additive manufacturing (AM) can be broadly illustrated as an instruction set representing a three-dimensional (3D) model sent from a computer to an AM system, through an interface [1–3]. This "standard" interface converts computer aided design

Standards, Quality Control, and Measurement Sciences in 3D Printing and Additive Manufacturing
http://dx.doi.org/10.1016/B978-0-12-813489-4.00004-0

(CAD) data of the 3D model into a format that is understood by the AM machine [4–6].

The origin of any AM fabricated parts stems from 3D model data, obtained from CAD systems or 3D scanners. After a 3D model is drafted or imported into a 3D CAD modeling program, such as SolidWorks, PTC Creo, Siemens NX, the model is then exported into various file formats, determined largely by the type of AM system used. Currently, there are more than 90 different file formats for 3D model software outputs [7]. Most file formats tend to be proprietary, which companies have developed for their own CAD programs. However, there are some generic formats that can be read by most CAD platforms. The more popular and widely used generic formats are:

- STL (stereolithography).
- IGES (initial graphics exchange specification).
- STEP (standardized graphic exchange format).
- OBJ (object file).
- VRML (virtual reality modeling language).
- NURBS (nonuniform rational basis spline).

Each of these formats has its own distinct advantages and disadvantages, which lead to some formats being more widely used than the others.

The flow of file formats usually follows the order illustrated in Fig. 4.1, starting with a file format output by the CAD software, usually a 3D parametric modeling platform, such as SolidWorks, and converted to STL format for printing on a desktop 3D printer. The slicing software slices the STL model into many different layers, and generates a set of toolpath

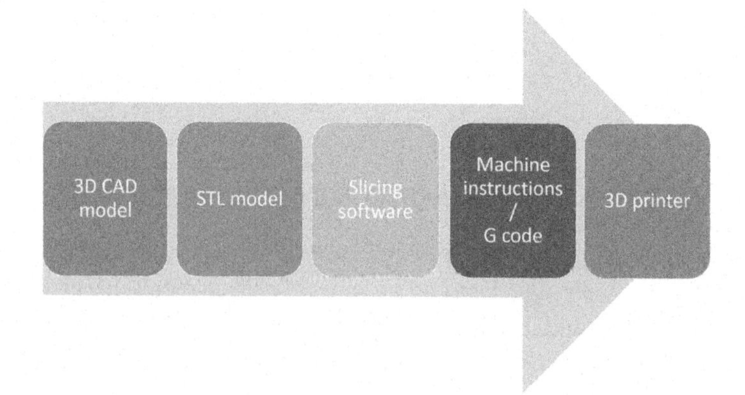

Figure 4.1 ***The Flow From a 3D Model to a Printed Part.***

instructions, such as G code for the 3D printer. The 3D printer, depending on the type of process used, refers to machine instructions or G code to execute the print, such as moving the printer head to a certain coordinate and begin the extrusion while moving the head to a new coordinate.

STL, originally developed by 3D Systems, is the present de facto file format that is accepted by most AM systems in the industry. Originally an acronym of STereoLithography, it was later attributed "backronyms," such as standard tessellation language or standard triangle language [8]. The wide adoption of STL file format in AM systems is due to its simplicity and independence from any CAD software. The file can be saved in 2 variants, binary and ASCII.

1.1 STL

STL file is a facet model derived from an approximation of a CAD model [9–12]. It is made of a surface mesh, which contains many sets of triangular facets, with each facet location described by a set of x, y, and z coordinates. The facet consists of three vertices and a unit normal vector to indicate the facet side within an object.

STL files may be in binary or ASCII format, where the latter is human-readable [13]. ASCII format allows the user to debug the file if there are any errors encountered but comes at the expense of a larger file size. In contrast, comparable binary STL files are smaller in size. For example, to store a numeric value of 10,000 in a computer, it would take about 6 bytes of storage space to store in ASCII compared to 4 bytes of storage space in binary.

STL file size increases with the number of facets. A high resolution model will generate a larger STL file than a low resolution model. Fig. 4.2 illustrates the difference in file sizes of a cylinder that was designed in a CAD software program, and then converted to an STL file for AM fabrication. To accurately represent the cylindrical shape, a high resolution STL

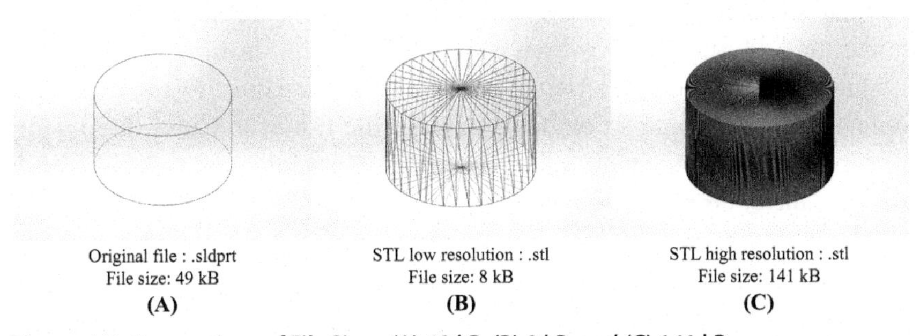

Figure 4.2 ***Comparison of File Sizes.*** (A) 49 kB, (B) 8 kB, and (C) 141 kB.

file would consist of numerous triangle facets to approximate the curved surface of the cylinder, which can be seen as a dense black cylinder on the right of the figure. On contrary, if the tolerance of the curvature is not crucial, a low resolution STL file can be used. This results in a reduction from 141 kB (high resolution) to 8 kB (low resolution).

1.2 STL File Problems

There are several problems associated with STL files due to their inherent nature, since they do not contain any topological data. Many commercial tessellation algorithms used by CAD vendors today are also not robust. As a result, they tend to create polygonal approximation models which exhibit the following types of errors:

- Missing facets or gaps.
- Degenerate facets (where all its edges are collinear).
- Overlapping facets.
- Nonmanifold topology conditions.

Early research has shown that repairing invalid models is difficult and the errors are not at all obvious [14,15]. The underlying problem is due to the difficulties encountered in tessellating trimmed surfaces, surface intersections and controlling numerical errors. This inability of the commercial tessellation algorithm to generate valid facet model tessellations makes it necessary to perform model validity checks before the tessellated model is sent to the AM equipment for manufacturing. If the tessellated model is invalid, repairing STL files become necessary to determine what the specific problems are, whether they are due to gaps, degenerated or overlapping facets, etc.

1.3 IGES

IGES is a data format used to exchange graphical information between CAD systems. The IGES file format, which was established as an American national standard in 1981, is able to accurately represent CAD models, unlike the STL file format which is merely an approximation.

There are a variety of AM systems, such as Stratasys 3D series, DTM Sinterstation 2000, and so on, that supports the use of IGES in their print software. IGES also contains information on points, lines, arc, curves, curved surfaces, and solid primitives to precisely represent CAD models.

On the contrary, as compared to STL, not all AM systems accept IGES files at present. Some disadvantages of IGES files include redundant information, lack of support for facet representation and complex algorithms.

1.4 STEP

STEP is a data format developed by ISO [16]. Published as ISO 10303, STEP covers a scope wider than many other CAD file formats. Compared to IGES and STL, the intention of STEP is to house more information, such as product related data that covers the entire life cycle of a product.

STEP is a result of a collaborative work involving several hundreds of people from different countries. It was first published in 1994 by members of the relevant standards subcommittee, ISO TC184/SC4. The STEP format is always expanding to cater to new requirements defined by the industry. It has the capability to contain information relating to materials, product life cycles, features, etc. STEP has about 40 different defined parts with many more in development [16]. Compared to IGES, STEP files can contain information, such as:

- Raw material data.
- Manufacturing tools.
- Manufacturing methods.
- Standard parts.
- Tolerance data.
- Features.
- Numerical control data.
- Data for kinematic simulation.
- 2D drawing.
- Product data management.

The most widely used part in STEP is the geometry data exchange, under application protocol AP 203 which deals with information regarding the shape of the product, assembly instructions and configuration instructions like part version, release status, etc.

Although STEP file format is widely used due to its nonproprietary nature, it was developed to cover the entire life cycle of the product from design to end-of-life. Hence, it contains a lot of data that may not be required for AM processes. In general, STEP requires new algorithms and interpreters to decipher the data for AM, similar to IGES [9].

1.5 OBJ

OBJ is an open source file format developed by Wavefront Technologies, originally for their Advanced Visualizer animation package [17]. The ASCII file format is used for 3D mesh geometries. OBJ is very similar to STL in both formats and the ability to support 3D meshes. However, OBJ has additional advantages by being able to support texture and material information, animation, and object hierarchy. Although it is universally accepted due to its simplicity, it is not widely supported by the community. There are some low-level entry printers that accept OBJ file on top of STL due to the similarity of 3D mesh language.

1.6 VRML

VRML is a standard file format that accommodates the following features [18]:
- Vertices and edges to represent a 3D polygon.
- Surface color.
- Transparency.
- Surface reflectivity.

It is used for representing 3D interactive vector graphics designed for the World Wide Web. However, the file format has been superseded by X3D, known as extensible 3D graphics [19].

These file formats enable users to visually see parts that have been uploaded online by others and subsequently download them for AM fabrication. To improve the usability, all files are automatically converted to STL format for download so that users can start printing right away.

1.7 NURBS

NURBS is a representation of a mathematical formula used to accurately model a curve or surface of an object [20]. NURBS can represent any form of shapes and surfaces that can be defined in 3D, such as straight lines, polynomial lines, free form curves,and so on.

The advantage of NURBS is the ability to represent complicated lines and shapes while keeping the overall file size small [21]. NURBS surfaces are smoother as they are also mathematically formulated, unlike STL where curves are attained by many facets and triangles formed together. This also greatly decreases the amount of data required to achieve similar results.

There are many different AM file format options for the industry. The six file formats stated previously are the more generic open formats generally accepted by all systems. However, as STL files lack the ability to carry more information, there are developments in defining newer file format standards specifically for AM. For example, ISO/TC 261 and ASTM F42 are working together to develop the AMF file format (additive manufacturing format) [22] and Microsoft is collaborating with industry partners to establish the 3MF consortium to develop the 3MF file format (3D manufacturing format) [23].

2 NEW DEVELOPMENTS IN DATA FORMATS

STL has been a de facto industry standard for AM since the early 1990s. Since then, there have been many other formats developed to solve the shortcomings of STL, but they largely failed to be widely adopted. However, to address the future needs of the AM industry, ASTM F42 and ISO/TC261 decided to jointly manage the AMF standard, while Microsoft, together with other partners, jointly formed the 3MF consortium [24].

AMF was proposed and implemented to address the needs of the AM community and to replace the ageing STL file format. STL can only contain mesh data, lacking the ability to carry information on attributes, such as material properties, color, etc. STL also has an issue with file size: if a round or spherical object is to be represented in STL format, the file size will increase exponentially according to the concentricity required. Furthermore, the computer may not have enough memory to print or handle STL files that are large in size. This creates a problem for printing as the user may not be able to open the file at all. Additionally, large STL files may contain many overlapping triangular facets that would be hard to repair. Therefore, the implementation of AMF will resolve some of these issues [25].

Although it is widely recognized that STL would continue to be important in the AM industry, a file format with more functionality is required. Microsoft, together with its partners, concluded that no such file format could meet the requirements and hence determined that the best approach would be to create a new 3D file format through a group effort with broad industry involvement. This gives rise to the birth of 3MF consortium and the 3MF file format, which is XML-based and includes information about materials, colors, and other information that cannot be represented in the STL format [26]. The 3MF file format is, however, only compatible with Microsoft Windows 8.1 and Windows 10 [27].

Similarly, many AM equipment manufacturers are also beginning to collaborate with software partners to allow their machines to print directly from any natively supported CAD format. For example, Stratasys is working with GrabCAD, an online CAD library sharing service, to push out

GrabCAD Print, which is expected to be compatible with most Stratasys machines [28].

Given the recent advancements in AM processes, there is an urgent need to overcome the limitations arising from the traditionally used file formats, such as STL. It is also essential to avoid fragmentation of AM file systems to ensure compatibility among different design and manufacturing platforms. The development and standardization of capable file formats are keys to faster adoption of AM processes in the industry.

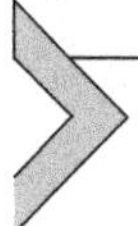

3 SCANNING TECHNOLOGY AND DATA FORMAT FOR ADDITIVE MANUFACTURING

In the context of AM, scanning enables possibilities in reverse engineering. A designer can scan an object of interest and replicate it through an AM system. For example, a designer can use a 3D scanner to generate a CAD model of the car that requires redesigning. Moreover, specific modifications can be added before finally printing the 3D model to obtain a physical replica of the modified CAD design. Compared to using measurement devices, such as micrometer and vernier caliper, the speed of scanning saves time while maintaining a relatively accurate replicate of the model. Using the scanner reduces the need to manually take physical measurements, saving time and effort. There are many technologies and variants behind the working principles of 3D scanners, and different manufacturers use different file formats.

Some examples of file formats generated by 3D scanners are [29]:

- .ply
- .obj
- .stl
- .aio
- .thing
- .off
- .wrl
- .aop
- .ascii
- .ptx
- .e57
- .xyzrgb
- .pl

Scanning technology can broadly be classified into contact and noncontact scanners [30]. Noncontact scanners, which are usually optical in nature, can be further classified into passive and active scanners.

3.1 Contact Scanners

A contact 3D scanner uses probes to communicate tactilely with the object to be scanned. These probes are usually mounted on a 3- or 5-axis machine, a robotic arm, or a combination of both. The mechanism behind contact scanners can be classified into three forms which are:

- A 3-axis system usually consists of 3 tracks held 90 degrees relative to each other to form a Cartesian positioning system. A contact probe is attached to this setup, to touch and move along the surface of the object of interest. These systems work best with flat profile shapes or simple curves and surfaces.
- A robotic articulated arm using angular sensors with a probe at the end of the arm. The arm rotates to measure the object using angular sensors and rotational sensors. The arm is capable of probing into interior spaces, convex shapes, and complex surfaces.
- A combination of the aforementioned two methods, usually used for large objects.

An example of a contact 3D scanner is a coordinate measuring machine (CMM). A carriage type CMM with three axes is widely used in the industry for measurement of parts to check for quality and consistency. CMMs usually use a ruby contact probe for measurement due to its stability [31]. However, there is a risk of damaging soft or polished parts during measurement due to physical contact. A CMM is relatively slow in nature as it can only measure one point at a time. Unlike the noncontact 3D scanners, it does point to point measurements and the surface is generated based on the average of all the points on that part.

3.2 Noncontact Scanners

Noncontact 3D scanners are similar to a camera in many ways. They capture images from a cone-like field of view and only can see surfaces that are not obstructed by any medium. However, unlike a camera, 3D scanners measure the distance from the surface to the scanner lens. Classified into active and passive, active noncontact scanners emit light or radiation (X-ray, gamma ray, etc.) to capture the reflection or radiation from an object to measure the part of interest. Unlike their active counterparts, passive noncontact scanners do not emit any form of light or radiation. Instead, the scanners rely on ambient light and reflection from surfaces for measurement. The technology behind noncontact scanners can be classified into one of the following forms:

Active

- Time-of-flight.
- Triangulation.
- Hand-held laser scanners.
- Structured light.
- Modulated light.
- Volumetric techniques like computed tomography.

Passive

- Stereoscopic system.
- Photometric system.
- Silhouette technique.

3.3 Time-of-Flight

Time-of-flight scanners usually emit pulses of laser or infra-red radiation, and these pulses are captured and measured by the receiver. The time taken for the round trip is the light travelling from the emitter and reflected back into the receiver. 3D scanning using a time-of-flight camera is relatively inexpensive compared to other specialized scanners and are able to produce relatively good scans [32].

3.4 Triangulation

A laser-based 3D scanner uses either a dot or stripe of laser light to measure and scan the 3D object. As the laser strikes on the different angles of a surface on an object, the difference in distance due to the reflection from the laser will be projected differently on parts of the camera sensor. As the information from the laser emission, camera, and the reflected laser on the camera are known, it is therefore possible to determine the shape and size of the triangle. One of the first developers of a triangulation based laser scanner was the National Research Council of Canada in 1978 [33].

3.5 Hand-Held Scanners

Hand-held scanners use triangulation methods to measure and scan a 3D object. They usually rely on reference points, such as white or black sticker dots that are placed on the object to track its relative position. The user has to move the scanner around the object to perform the scan. Some handheld scanners utilize infra-red light while others make use of white light for the scans. Newer generation of hand-held scanners have in-built gyroscopes and accelerometers to track their position relative to the object.

3.6 Structured Light

Structured light 3D scanners utilize the rapid shifting deformation of light patterns emitted onto a surface to determine the shape of an object. These sources are usually either blue or white light. A camera captures the patterns projected onto the object, measures the deformations of the patterns, and translates it into a 3D model on the computer. As this technology is relatively new, there is much ongoing research to improve the accuracy of the structured light [34]. Research is also conducted on the shapes and patterns of the light source, such as vertical or horizontal moving stripes, dots and even grids, to scan the object [35]. Ambient lighting may introduce noise during the scan and therefore it is important that the ambient lights are dimmed so as not to affect the scan quality [36]. Certain scanners utilize blue instead of white or yellow light to reduce the noise during scanning caused by the ambient light [37]. Advantages of structured light scanners include high speed and high precision scanning, due to its ability to scan an entire area field as opposed to laser based scanning where only one thin line is scanned at a time as it slowly moves over the entire object [38]. The precision and speed depend on the type of cameras used for the scanning.

3.7 Modulated Light

Modulated light 3D scanners emit a fluctuating light source at the object of interest [39]. By dimming or brightening the light source using a sine function, the camera will be able to detect the light pattern shift casted on the object. These changes in the light shift allow the camera to determine points on the object and its relative distance to the camera, thus forming a 3D model. Modulated light scanning is not affected by changes in ambient light.

3.8 Volumetric Method

Volumetric scanning uses stacked images from thousands of images scanned throughout an object. Computed tomography (CT) scan is one such popular method widely used in the medical industry. This will be further discussed in Section 4.

Unlike conventional light and laser scanning, a CT scan can also be used to scan internal structures of molds, cooling channels, etc. As a CT scan only provides 2D cross-sectional images, a software package may therefore be used to convert the cross-sectional images into a 3D model. The advantages of CT scanning are the ability to detect voids, channels, and passages within an object that is not possible by surface scanning.

3.9 Stereoscopic System

Working on a principle similar to human eyes, a stereoscopic system uses two cameras that are placed slightly apart from each other. By analyzing the slight differences in the images, the system is able to measure the distance from the camera to the object, and thus generate a surface profile of that object [39]. Comparing the images captured by the cameras, greater offset is observed between points on an object that are nearer to the cameras, as compared to those further away. As the distance between the two cameras is fixed and known, the distance from the cameras to the individual points on the object can be trigonometrically calculated.

3.10 Photometric System

A photometric system uses a single camera setup, unlike the dual cameras in a stereoscopic system. To measure the profile of an object, it captures multiple images under different lighting conditions and then inverts the image model to obtain information on the surface at each pixel on the camera sensor [39]. The model formed will be based on different images at different angles. Any area not captured by the camera will not have any surface generated and hence resulting in a void in the image.

3.11 Silhouette Techniques

Silhouette techniques utilize silhouette projections that are taken against a background to form an approximation of the scanned part [39]. The camera will be able to capture an overview of the shape or size of the object, although concave features cannot be detected. Usually, on a rotating base, the different 2D images that are taken by the camera will be combined to form a 3D model.

3.12 Point Cloud

Most 3D scanners output raw scanned data in point cloud format. Point clouds produced by 3D scanners and 3D imaging are visualized for the ease of measurement. A point cloud is basically a set of data points in a 3D coordinate system, commonly defined by x, y, and z coordinates. They are used to represent the surface of an object and do not contain data of any internal features, color, materials, and so on. However, many CAD applications employ parametric, direct, or mesh-based modeling. Therefore, there is a need to convert the data from the point cloud into formats natively adaptable to CAD software in order to obtain geometric information of the scanned part. The conversion of point cloud data to STL files also allows for ease of fabrication using AM [40].

Parametric modeling uses a feature-based approach to design the shape of the model. Features can be elements, such as bosses, holes, fillets, chamfers, cuts, which when combined with the model form the final part [41]. Furthermore, parametric modeling is history-based, making it possible to select and edit certain features in the history tree if required. However, if a particular feature is referenced by any other features created in the future, deleting or changing this referenced feature will affect the future features. Parametric modeling is the most common form of modeling used, and point cloud data can be converted into a parametric model for editing, before exporting it to a file format suitable for AM.

Direct modeling, similar to synchronous technology by Siemens and direct modeling by PTC, is a nonhistory based direct editing technique for creating 3D models [42,43]. Unlike parametric modeling, direct modeling allows editing of a feature with real-time visualization. The designer can shift, for example, the location of a boss without the need of any numbers from one location to another, and any relationship the boss has with any other features will be automatically updated real time. Direct modeling is relatively faster compared to parametric modeling.

Mesh-based modeling, also known as polygonal modeling, uses polygons, usually a triangle, as the smallest element in the modeling. Depending on the configuration, it can be made into a surface or 3D model. The point cloud can be easily converted to a mesh-based model as the points can be easily connected to form a mesh. With a larger number of points, the mesh becomes very fine and more computational power is required to handle the file. Although the STL file format is also based on the same modeling technique, by importing the point cloud to STL allows direct AM printing provided that the scan is of high resolution and no repairs on the file is required.

Regardless of what file format the point cloud is converted into, it is crucial that the object scanned is well represented in the software before the build begins. If necessary, amendments should be made before sending it to other applications, platforms, services, or AM equipment.

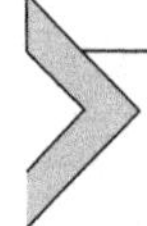

4 MEDICAL IMAGING AND CONVERSION SOFTWARE FOR ADDITIVE MANUFACTURING

Images for medical diagnostics applications are fundamentally sourced by means of X-rays or magnetic resonance. Data for the construction of 3D models is normally obtained by fluoroscopy or CT scanners that work on the principle of X-ray radiography, or magnetic resonance imaging (MRI) scanners that obtain images by magnetic resonance.

Data obtained from a CT or MRI scan consists of a combination of many cross-sectional 2D images taken from different angles. Digital geometry processing is used to generate a 3D image of an inside of the object from a series of 2D radiographic images taken around a single axis of rotation [44].

While most CAD software possesses some basic surfacing and reverse engineering capabilities, the complexity involved with surfaces relating to medical applications requires well designed algorithms to handle the highly uneven and organic nature of surfaces. In addition to being able to work with point cloud data derived from external scanners that work on light triangulation, the medical imaging and conversion software should be able to work directly with native files obtained from standard medical diagnostic equipment, such as CT and MRI scanners.

Digital imaging and communications in medicine (DICOM) is a standard that specifies a file format, as well as a network communications protocol for medical imaging. The DICOM file format is widely used by manufacturers of medical imaging equipment, information systems and other peripheral equipment in the medical industry [45]. The software also needs to possess capabilities to identify, classify and filter different constituents in the medical scan, such as bone and soft tissue. This is done by a process broadly termed as segmentation, where the contour of a particular geometry, such as a bone segment, is identified at different cross-sectional layers of a CT or MRI scan, and the software is able to isolate and form a 3D model of this identified geometry. Any imaging and conversion software focused on AM for medical applications is able to natively import and process DICOM files, allow for segmentation and generate 3D models. These 3D models can then be exported to standard AM formats, such as STL or AMF.

4.1 Mimics

Mimics, a software package from Materialise NV, imports DICOM scans, processes them and forms accurate 3D models that can be used for analyses. The software focuses on the following six applications [46]:

- Medical image segmentation.
- Anatomical analysis.
- Virtual surgery.
- Benchtop model design.
- Patient-specific device design.
- Postoperative analysis.

The software allows surgeons to assess their patients with a better understanding of their bodies. As each patient is unique, customization of medical devices, surgical tools and mounts, and even virtual surgery are made possible through software advancement.

5 SOFTWARE AND DATA VALIDATION

All scanned data will have some form of accuracy issues due to noise. However, most noise is filtered during the scanning process. To ensure that a part is well scanned, the user has to check if there are repairs needed to smoothen the part model, or if any part needs rescanning. Print quality can be improved by utilizing different slicing techniques that are used to dissect the STL file into different layers for AM process. Fig. 4.3 depicts a fully sliced part with printer's toolpath.

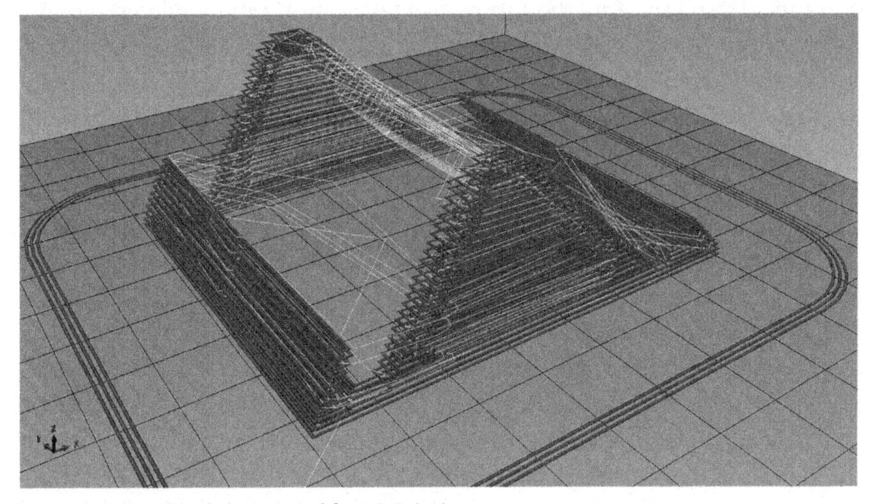

Figure 4.3 ***Toolpath Generated for 3D Printing.***

A method to improve the quality of a model that has been scanned, is to smoothen the STL model by using Max-Fit biarc curves technique. As mentioned in Section 2, STL files require more storage space if a better representation of a curve is desired. Therefore the density of triangles used in the STL has to be increased to generate a more accurate curve, which results in a large file size [47]. The STL model will then be sliced into layers required for the AM process, and the sliced data will be sent to the AM machine for fabrication. Since curves for AM processes are usually generated by many small straight line segments joined to form a curve, the surface finish is usually inferior. These straight line segments in the sliced toolpath data are processed likewise by the machine. However, if the sliced toolpath data can contain curves, the surface quality will be smoother when the part is printed out, as the print head will move in a continuous curve rather than a series of many straight vector lines. The Max-Fit biarc curve algorithm can be implemented during STL slicing to fit a close curve while maintaining the tolerances required to improve the accuracy of the print and at the same time keeping the STL file size small [47].

Another technique to improve print quality is to utilize adaptive slicing. There are many approaches using adaptive slicing for AM. Three adaptive slicing approaches will be discussed: local adaptive slicing, stepwise uniform refinement and cubic patch approximation.

Typical slicing methods depend on the layer thickness required. A thicker slice will result in a poorer surface finish on a curved surface. This is due to staircase like effects that are generated during slicing, which can be seen in Fig. 4.4. However, if the slices are too thin, it will result in a longer fabrication time. Therefore, a method to reduce fabrication time while preserving good surface finish is to employ local adaptive slicing [48]. It is proposed that parts with simple surfaces can utilize thickly sliced layers while parts with complex surfaces use thinly sliced layers [48]. It is possible to apply this slicing technique to those AM processes with the ability to print different layer thickness throughout the build.

Unfortunately, local adaptive slicing cannot be universally applied to all AM processes since some AM systems only allow a fixed slice parameter. For example, an extrusion nozzle of FDM machine has a fixed nozzle diameter, which determines the extrusion diameter and the thickness of the layer. In order to produce different layer thicknesses, a variable nozzle diameter would then be required.

Local adaptive slicing, however, can be applied to powder bed processes. If a particular set of layers only contain simple surfaces, thicker recoating

Figure 4.4 ***Staircase Like Effect From Slicing.***

to increase the layer thickness is possible. The laser power can also be proportionately increased to sinter a thicker layer. If another set of layers contains complex surfaces, each layer can be recoated thinly and sintered with reduced laser power. These thin layers, over many layers, can achieve complex surfaces with reduced staircase-like effects.

Stepwise uniform refinement is another technique used to initially divide the model into thick slices. Those thick slices that have complex surfaces on them will be further subdivided individually to meet the maximum cusp height in order to achieve the best possible surface finish [49]. However, it is important that the user recognizes the critical features in the model when using this technique. Thicker slices may also lead to loss of fine complex features. Hence, it may be necessary to consider additional slice layers for regions with such features during the slicing process [49]. Thus with stepwise uniform refinement, complex features can still be produced while saving time on less complex areas that are sliced for speed.

In cubic patch approximation, the slice of an outer wall, which can be considered as the external surface, can be modelled as a pair of successive contours instead of straight lines generated from the STL triangular facets

[50]. Kumar and Choudhury suggested that a Bezier surface can be used by defining the corner points of a cubic patch which are divided into grids. The grid points that are normal to the surface are measured and the deviation between all the normals is used as the cusp height for the portion of the outer wall. Therefore a curvature is generated from the slices resulting in a better surface finish.

The previously described slicing techniques can be used to greatly speed up the AM process, but it is crucial to check and ensure that any required features of the model are not lost during the slicing process [49].

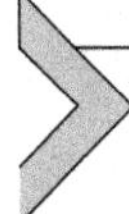

6 QUESTIONS

1. Starting from an existing physical model, list the steps required to print a replica of this model part using an AM machine.
2. List the limitations of the STL file format. What are the solutions that have been developed to address these limitations?
3. What are the types of 3D scanning technology available? List their advantages and disadvantages.
4. Why is there a need for DICOM?
5. What are the steps involved in generating a 3D model of an internal organ?
6. What techniques can be adopted to improve printing speed and surface finish while ensuring that critical features are printed?

REFERENCES

[1] C.K. Chua, K.F. Leong, 3D Printing and Additive Manufacturing: Principles and Applications, fifth ed., World Scientific Publishing Company, Singapore, (2017).
[2] C.K. Chua, M.V. Matham, Y.J. Kim, Lasers in 3D Printing and Manufacturing, World Scientific Publishing Company, Singapore, (2017).
[3] C.K. Chua, W.Y. Yeong, Bioprinting: Principles and Applications, World Scientific Publishing Company, Singapore, (2014).
[4] Y.K. Modi, S. Agrawal, D.J. de Beer, Direct generation of STL files from USGS DEM data for additive manufacturing of terrain models, Virtual Phys. Prototyp. 10 (2015) 137–148.
[5] A. Ghazanfari, W. Li, M.C. Leu, Adaptive rastering algorithm for freeform extrusion fabrication processes, Virtual Phys. Prototyp. 10 (2015) 163–172.
[6] J.-Y. Lee, W.S. Tan, J. An, C.K. Chua, C.Y. Tang, A.G. Fane, T.H. Chong, The potential to enhance membrane module design with 3D printing technology, J. Memb. Sci. 499 (2016) 480–490.
[7] S. Tibbits, 4D printing: multi-material shape change, Architect. Design 84 (2014) 116–121.

[8] 3D Systems Inc., The STL format, in: StereoLithography Interface Specification, 1989.
[9] C.K. Chua, G.K.J. Gan, M. Tong, Interface between CAD and Rapid Prototyping systems. Part 1: a study of existing interfaces, Int. J. Adv. Manufactur. Technol. 13 (1997) 566–570.
[10] C.K. Chua, G.K.J. Gan, M. Tong, Interface between CAD and Rapid Prototyping systems. Part 2: LMI — An improved interface, Int. J. Adv. Manufactur. Technol. 13 (1997) 571–576.
[11] G.K.J. Gan, C.K. Chua, M. Tong, Development of a new rapid prototyping interface, Comput. Ind. 39 (1999) 61–70.
[12] C.K. Chua, K.F. Leong, 3D Printing and Additive Manufacturing: Principles and Applications, fourth ed., World Scientific Publishing Company, Singapore, (2014).
[13] C.K. Chua, K.F. Leong, C.S. Lim, Rapid Prototyping: Principles and Applications, 3rd ed., World Scientific Publishing Company, Singapore, (2010).
[14] K.F. Leong, C.K. Chua, Y.M. Ng, A study of stereolithography file errors and repair. Part 1. Generic solution, Int. J. Adv. Manufactur. Technol. 12 (1996) 407–414.
[15] K.F. Leong, C.K. Chua, Y.M. Ng, A study of stereolithography file errors and repair. Part 2. Special cases, Int. J. Adv. Manufactur. Technol. 12 (1996) 415–422.
[16] M.J. Pratt, Introduction to ISO 10303—the STEP standard for product data exchange, J. Comput. Inform. Sci. Eng. 1 (2001) 102–103.
[17] OBJ files a 3D object format (2017). Available from: http://people.sc.fsu.edu/~jburkardt/data/obj/obj.html
[18] D. Raviv, W. Zhao, C. McKnelly, A. Papadopoulou, A. Kadambi, B. Shi, et al. Active printed materials for complex self-evolving deformations, Sci. Rep. 4 (2014) 7422.
[19] X3D recommended standards (2017). Available from: http://www.web3d.org/standards
[20] P. Lavoie, An introduction to NURBS, Philippe Laovie, 1998. Avaialable from: http://www.hcs.harvard.edu/~lynders/cs275/nurbsintro.pdf.
[21] P.J. Schneider. (2017). NURB curves: a guide for the uninitiated. Available from: http://www.mactech.com/articles/develop/issue_25/schneider.html
[22] ISO and ASTM, Standard specification for additive manufacturing file format (AMF) version 1.2, in: ISO / ASTM52915-16, ed: ISO, ASTM International, 2013.
[23] S. Badalov, Y. Oren, C.J. Arnusch, Ink-jet printing assisted fabrication of patterned thin film composite membranes, J. Membr. Sci. 493 (2015) 508–514.
[24] S. Badalov, C.J. Arnusch, Ink-jet printing assisted fabrication of thin film composite membranes, J. Membr. Sci. 515 (2016) 79–85.
[25] P. Kinnane. (2017). A better file format for 3D printing to replace STL. Available from: https://www.comsol.com/blogs/a-better-file-format-for-3d-printing-to-replace-stl/
[26] D. Bella. (2015). 3D printing file format cage match: AMF vs. 3MF. Available from: http://blog.grabcad.com/blog/2015/07/21/amf-vs-3mf/
[27] Y. Lu, S. Choi, P. Witherell, Towards an integrated data schema design for additive manufacturing: conceptual modeling in: ASME 2015 International Design Engineering Technical Conferences and Computers and Information in Engineering Conference, Boston, Massachusetts, USA, 2015.
[28] Grabcad print (2017). Available from: https://grabcad.com/print
[29] D. Kim, S.H. Lee, S. Jeong, J. Moon, All-ink-jet printed flexible organic thin-film transistors on plastic substrates, Electrochem. Solid State Lett. 12 (2009) H195–H197.
[30] B. Curless, From range scans to 3D models, J. Comput. Graph. (ACM) 33 (4) (1999) 38–41.
[31] At the sharp end - a guide to CMM stylus selection (2017). Available from: http://www.renishaw.com/en/at-the-sharp-end-a-guide-to-cmm-stylus-selection--10927
[32] Y. Cui, S. Schuon, D. Chan, S. Thrun, C. Theobalt, 3D shape scanning with a time-of-flight camera in: 2009 IEEE Conference on Computer Vision and Pattern Recognition (CVPR), Miami, FL, USA, 2010, pp. 1173–1180.
[33] R. Mayer, Scientific Canadian: invention and innovation from Canada's national research council, Raincoast Books, Vancouver, 1999.

[34] R.J. Valkenburg, A.M. McIvor, Accurate 3D measurement using a structured light system, Image Vision Comput. 16 (1998) 99–110.
[35] J. Geng, Structured-light 3D surface imaging: a tutorial, Adv. Optics Photon. 3 (2011) 128–160.
[36] M. Gupta, A. Agrawal, A. Veeraraghavan, S.G. Narasimhan, Structured light 3D scanning in the presence of global illumination, in: IEEE CVPR 2011 Conference, Colorado Springs, 2011, pp. 713–720.
[37] Blue light 3D scanners (2016). Available from: http://www.capture3d.com/3d-metrology-solutions/3d-scanners
[38] I. Ishii, K. Yamamoto, K. Doi, T. Tsuji, High-speed 3D image acquisition using coded structured light projection, in: 2007 IEEE/RSJ International Conference on Intelligent Robots and Systems, San Diego, CA, USA, 2007, pp. 925–930.
[39] M.A.-B. Ebrahim, 3D laser scanners' techniques overview, Int. J. Sci. Res. 4 (2015) 323–331.
[40] P. Pal, An easy rapid prototyping technique with point cloud data, Rapid Prototyp. J. 7 (2001) 82–90.
[41] R. Janusziewicz, J.R. Tumbleston, A.L. Quintanilla, S.J. Mecham, J.M. DeSimone, Layerless fabrication with continuous liquid interface production, Proc. Natl. Acad. Sci. 113 (2016) 11703–11708.
[42] J.-Y. Lee, J. An, C.K. Chua, Fundamentals and applications of 3D printing for novel materials, Appl. Mater. Today 7 (2017) 120–133.
[43] M. Brunelli. (2016). Parametric vs. direct modeling: Which side are you on? Available from: http://www.ptc.com/cad-software-blog/parametric-vs-direct-modeling-which-side-are-you-on
[44] G.T. Herman, Fundamentals of Computerized Tomography: Image Reconstruction from Projections, second ed., Springer-Verlag, London, (2009).
[45] NEMA, Digital imaging and communications in medicine (DICOM) standard, PS3 / ISO 12052, National Electrical Manufacturers Association, Rosslyn, VA, USA. Available from: http://medical.nema.org/).
[46] I. Smurov, M. Doubenskaia, A. Zaitsev, Comprehensive analysis of laser cladding by means of optical diagnostics and numerical simulation, Surf. Coating. Technol. 220 (2013) 112–121.
[47] B. Koc, Y. Ma, Y.-S. Lee, Smoothing STL files by max-fit biarc curves for rapid prototyping, Rapid Prototyp. J. 6 (2000) 186–205.
[48] J. Tyberg, J.H. Bøhn, Local adaptive slicing, Rapid Prototyp. J. 4 (1998) 118–127.
[49] E. Sabourin, S.A. Houser, J.H. Bøhn, Adaptive slicing using stepwise uniform refinement, Rapid Prototyp. J. 2 (1996) 20–26.
[50] M. Kumar, A.R. Choudhury, Adaptive slicing with cubic patch approximation, Rapid Prototyp. J. 8 (2002) 224–232.

CHAPTER FIVE

Material Characterization for Additive Manufacturing

Contents

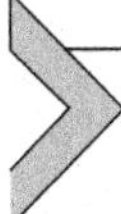

1 INTRODUCTION TO MATERIALS CHARACTERIZATION IN ADDITIVE MANUFACTURING

The typical materials used in AM currently are polymers and metals, with ongoing research expanding the usage to other groups of materials such as ceramics and composites. The raw materials used in AM can be broadly classified based on their forms in, either solid, liquid, or powder. The

Standards, Quality Control, and Measurement Sciences in 3D Printing and Additive Manufacturing
http://dx.doi.org/10.1016/B978-0-12-813489-4.00005-2

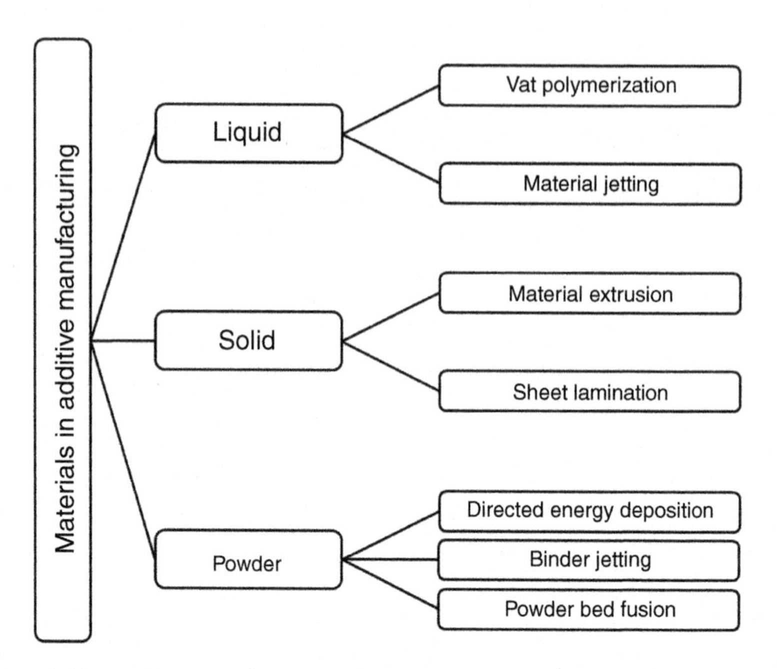

Figure 5.1 ***Types of Material in AM and Their Corresponding Process Categories.***

different types of material in AM and their corresponding process categories in ASTM/ISO standards are shown in Fig. 5.1.

1.1 Liquid-Based Additive Manufacturing

The representative liquid-based AM techniques include vat polymerization and material jetting. In vat polymerization, an ultraviolet (UV) light is used to cure or harden the photopolymer resin in a vat where required, while a platform moves the object to be printed upward or downward after each new layer is cured. Examples of such processes include stereolithography (SL) and digital light processing (DLP). In material jetting, droplets of the build material are dispensed selectively. These droplets are then cured by UV light. Example of such processes includes inkjet printing. Schematics of liquid-based AM techniques are shown in Fig. 5.2.

1.2 Solid-Based Additive Manufacturing

The most commonly seen technique in solid-based AM techniques is material extrusion, where materials, usually in the form of filament, are dispensed selectively through a nozzle or orifice. An example of such techniques is fused deposition modeling (FDM), as shown in Fig. 5.3.

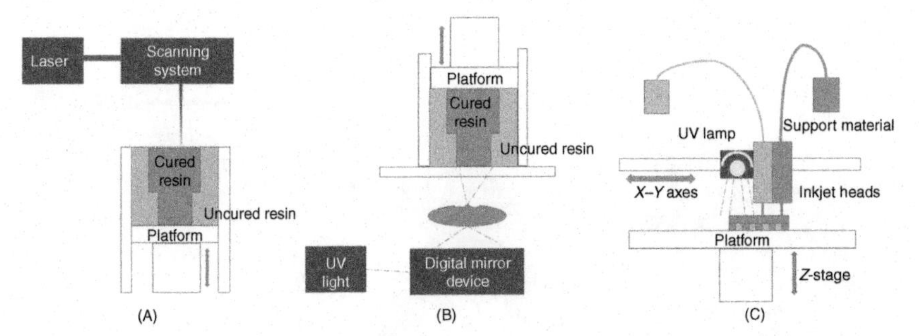

Figure 5.2 ***Liquid-based additive manufacturing (A) stereolithography (B) digital light processing (C) inkjet printing.***

Another solid-based AM technique, sheet lamination, also known as laminated object manufacturing (LOM), bonds sheets of materials together to form the part.

1.3 Powder-Based Additive Manufacturing

Powder-based AM techniques can be represented by directed energy deposition (DED), binder jetting and powder bed fusion (PBF). In DED, focused thermal energy, such as laser, electron beam or plasma arc, is used to fuse the material by melting as they are being deposited. An example of such processes is laser engineered net shaping (LENS). In binder jetting, a liquid

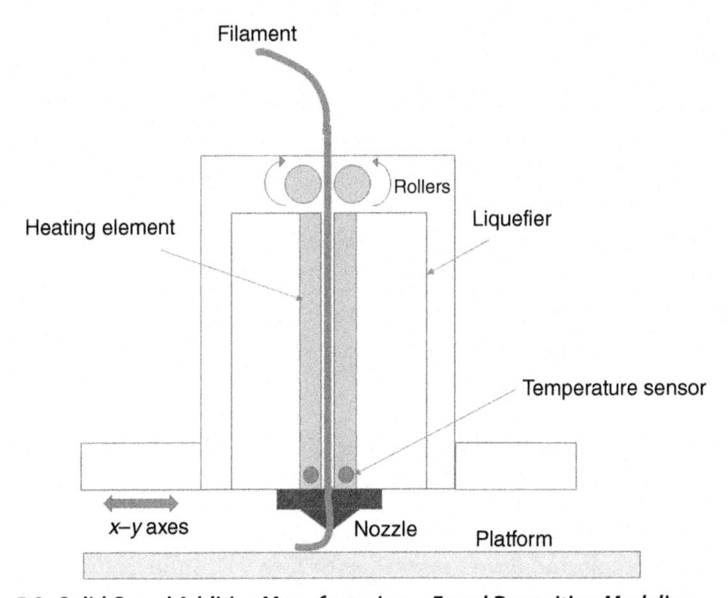

Figure 5.3 ***Solid-Based Additive Manufacturing—Fused Deposition Modeling.***

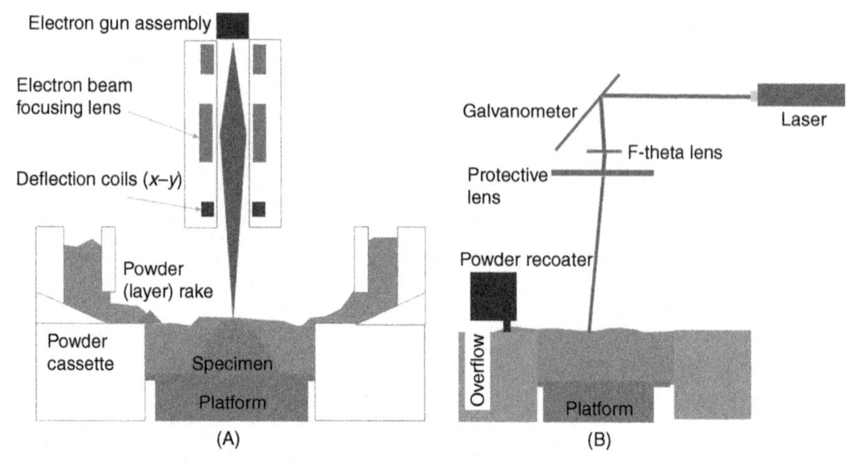

Figure 5.4 Powder-based additive manufacturing (A) electron beam melting (B) selective laser melting.

bonding agent is deposited selectively and used to join powder materials. In PBF techniques, thermal energy is used to selectively fuse regions of a powder bed, such as in selective laser melting (SLM) and electron beam melting (EBM). Schematics of SLM and EBM are shown in Fig. 5.4.

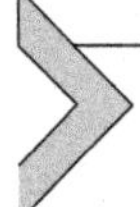

2 LIQUID MATERIALS CHARACTERIZATION TECHNIQUES

This section discusses on some important characterization techniques used for characterizing liquid materials. The list further shows some of the tests that will be discussed in the following subsections:

- viscosity measurement techniques,
- surface tension measurement techniques,
- curing characterization techniques,
- thermal stability assessment techniques,
- appearance assessment techniques, and
- density measurement technique.

2.1 Rheology and Wetting Behavior

2.1.1 Viscosity

Viscosity is a property of the fluid which resists the relative motion within the fluid, or simply, the fluid's resistance to flow [1]. Informally, the "thickness" of a liquid is commonly used to represent the liquid's viscosity. In fluid dynamic, viscosity of a fluid is a measure of its resistance to the gradual deformation of shear stress or tensile stress [1]. The international system of units, also known as SI unit for viscosity, is Pascal second (Pa·s), but the most common unit used for viscosity is dyne second per square centimeter (dyne·s/cm^2), or simply Poise (P). Mathematically, 1 Pa s is equivalent to 10 P, which makes centi-Poise and milli-pascal second identical.

In liquid-based processes, the raw materials are often liquid resin. Hence, the viscosity of the liquid resin plays an important role in determining its compatibility with the printing processes. In the case of top-down stereolithography, high viscosity resin contributes to leveling issue during printing and shrinkage of printed parts [2]. Without the wiper, resin with high viscosity requires more time to spread uniformly over the build platform to supply material for each layer. Hence, low viscosity resin is preferred to reduce the resin leveling time [2].

Similarly, for inkjet printing, the ability of the resin to form droplet and jetting is very much dependant on both the force of the actuator and the resin's viscosity [3]. Typically, high viscosity resin requires higher actuating force to be jetted through the nozzle. In other words, resin with too high viscosity would resist the flow through the nozzle to an extent that cannot be overcome by the actuator.

As such, each liquid-based process has a recommended viscosity range for optimum operation. The recommended viscosity level for inkjet printing resin at room temperature and dispensing temperature is 20–125 cP [4]. Resins used in stereolithography have higher viscosity range, typically between 90 and 2500 cP at 30°C, and for DLP resins, 50 to 1200 cP [5].

The characterization techniques that can be used to measure viscosity are:

- capillary viscometers, and
- orifice/cup viscometers.

Capillary viscometers are typically used to determine the viscosity of Newtonian fluid such as liquid resin. The viscosity of resin is determined by measuring the time taken for a quantity of resin to flow through a capillary of known diameter and length [6]. For laminar flow of resin, the kinematic viscosity of the resin is linearly proportional to the measured time by a capillary factor. The driving force the resin flow is the gravitational pull. The advantage of using gravitational pull as driving force is the high reliability, whereas the limitation is that it is not suitable for highly viscous samples. There are many variants of standardized capillary tube in use and they can be categorized as direct flow and reverse flow capillaries. Direct flow capillaries, such as Ostwald, Ubbelohde, and Cannon-Fenske capillaries, have reservoirs located below the measuring marks whereas reverse

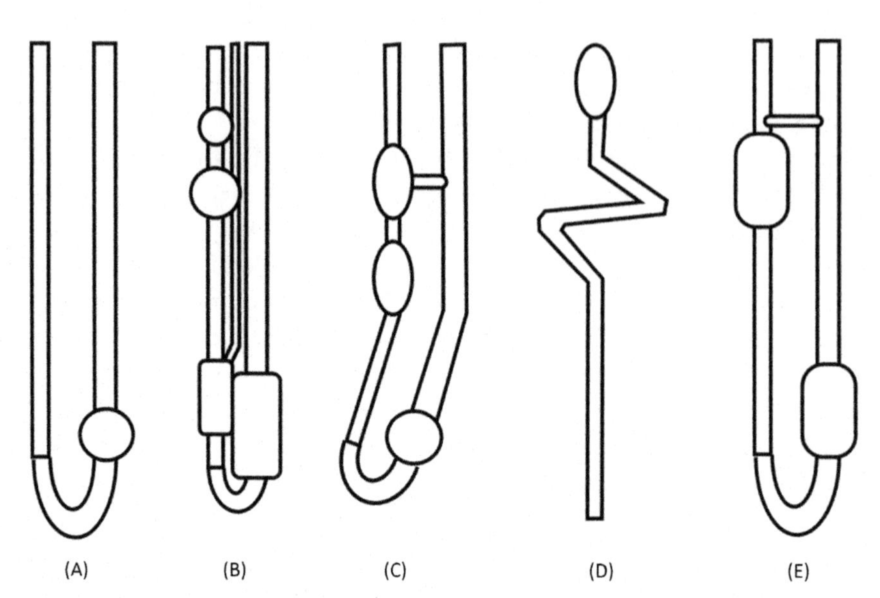

Figure 5.5 ***Examples of Capillary Viscometers.*** (A) Ostwald (B) Ubbelohde (C) Cannon-Fenske (D) Houillon (E) BS/IP/U tube

flow capillaries, such as Houillon and BS/IP/U tube, have reservoirs located previously the measuring marks while reverse flow usually has a third measuring mark which improves measurement repeatability and allow the use of opaque liquid [7]. Fig. 5.5 shows some example of the capillaries.

ASTM D446 provide a detail description of the operating instructions when using viscometer. The standards also discussed about the viscometer alignment, viscometer calibration and viscosity calculation

ASTM D446 Standard specifications and operating instructions for glass capillary kinematic viscometers.

This standard provides the specification, instructions, calibration procedures and basic calculation formulae for glass capillary viscometer widely used for determination of kinematic viscosity. It covers Ostwald viscometer, suspended level viscometer, and reverse-flow viscometer.

Similar to capillary viscometer, orifice viscometer uses gravitational force to pull the resin through an orifice located at the bottom of the viscometer. ASTM D4212-16 provides a detail description of the test procedures for viscosity measurement using cup viscometer.

ASTM D4212-16 Standard test method for viscosity by dip-type cups

This test method provides information about viscosity determination of paints, varnishes, lacquers, inks, and related liquid materials using

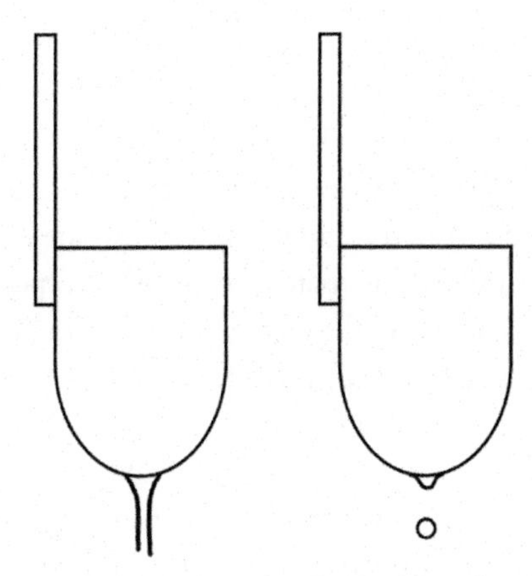

Figure 5.6 ***Continuous Flow, Changing Into Individual Droplets in an Orifice Viscometer.***

dip-type viscosity cups. This is suitable for testing Newtonian and near-Newtonian liquids. This test method is suitable for viscosity control work within a plant or laboratory.

This is a simple technique, which requires only an orifice viscometer, thermometer, and a stopwatch. The viscometer is first immersed in the liquid resin to fill the cup. Some precaution to take note is to ensure no bubble is formed when immersing the cup and make sure that the resin temperature in the cup is the same as the vat temperature. The cup is then slowly lifted off the resin surface and the stopwatch is started once the cup leaves the surface. The stopwatch is stopped once the stream of resin breaks and changes from continuous to individual drops. According to ASTM D4212-16, viscosity measurements have to be repeated minimum two times and difference between results should be smaller than 11%. A schematic of continuous flow changing to individual stops is shown in Fig. 5.6.

2.1.2 Surface Tension

Surface tension of a liquid is a result of imbalance between the intermolecular attractive forces (cohesive forces) of the liquid molecules at its interface. Unlike other molecules within the liquid, the molecules at the interface do not have other similar molecules on all sides resulting in a net inward pulling force which tends to pull the molecules back into bulk liquid, leaving minimum number of molecules at its interface. In thermodynamics, surface tension is measured as the work required changing the size of a liquid surface by a unit area. The SI unit for surface tension is Newton per meter; however, centimeter–gram–second unit (dyne per centimeter) is more commonly used instead.

Surface tension of resin in liquid-based systems plays an important part in determining its printability for various processes. For both SL and inkjet printing, the surface tension of resin determines it's wetting behavior on the substrates. The wetting behavior is expected to be better when the surface tension of the substrate is much higher than the surface tension of resin [5]. It is especially crucial for polymeric materials like ABS as polymers tend to have a low surface tension, which could result in poor wetting behavior. Apart from affecting wetting behavior, the surface tension of resin affects the jetting of the resin through the nozzle in inkjet printing technique. If the resin surface tension were too high, the resin would not be able to be jetted through the nozzle by the actuator. However, if the resin surface tension is too low, a splash of fluid will exit the nozzle instead of a cohesive droplet, which is essential for good printing. Furthermore, the surface tension of resin also determines the droplet size coming out from the nozzle. For optimum operation, the recommended surface tension for inkjet printing technique is around 30 dyne/cm, whereas the recommended surface tension for SL is 30 to 42 dyne/cm [5]. The surface tension of the resin can be reduced by adding a surfactant to the resin formulation to improve jetting.

The characterization techniques that can be used to measure surface tension are [8]:

- capillary rise method,
- stallagmometer method, and
- wilhelmy plate method.

As its name implies, capillary rise method uses the phenomenon of rising/dropping liquid column in a thin tube to determine the surface tension. This is one of the oldest techniques for surface tension measurement. Normally, the thin tube is made of glass. For accurate surface tension measurement, the diameter of the thin tube has to be sufficiently small to ensure hemispherical concave/convex meniscus at the liquid–air interface. Then, the surface tension γ can be determined by using the equation below:

$$\gamma = \frac{rh\rho g}{2\cos\theta}$$

where, r is the radius of the thin tube; h is the rise/drop in liquid column; ρ is the density of resin; g is the gravity pull and θ is the contact angle. A schematic of the capillary rise method is shown in Fig. 5.7.

Stallagmometer method, also known as drop weight method, measures the surface tension by weighing the drops coming out from the tip, as shown in Fig. 5.8. The fundamental principle of this method is based on the

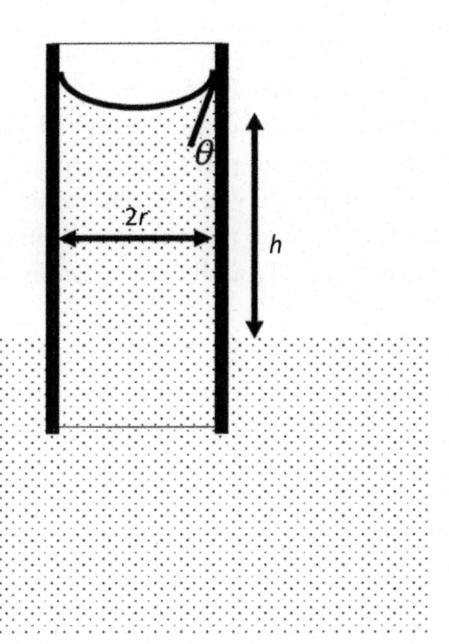

Figure 5.7 ***Schematic Representation of the Capillary Rise Method.***

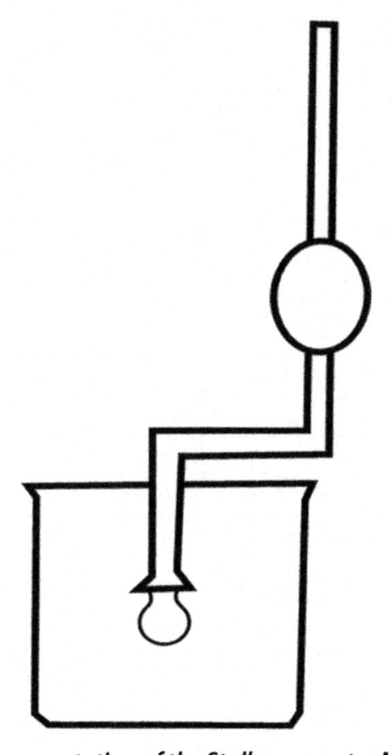

Figure 5.8 ***Schematic Representation of the Stallagmometer Method.***

force equilibrium of the surface tension and the weight of the drops. The pendant drops start to detach from the tip when the weight of the drops reach the magnitude of the surface tension. Ideally, the drops should fall off completely from the tip but part of the drops is retained in reality. Hence, a correction factor is normally introduced to the calculation of surface tension. It is important to note that the correction factor is dependent on the liquid resin being tested. Typically, the equation below is used:

$$\gamma = \frac{mg}{2\pi rf}$$

where, m is the drop weight; g is the gravitational pull; r is the radius of the drop and f is the correction factor. A schematic of the Stallagmometer method is shown in Fig. 5.8.

Wilhelmy plate method can be used to measure surface tension of liquid, contact angle between liquid and solid. Typically, a clean glass/aluminum plate with weight, W and width, L is used to measure the liquid–solid interface tension and liquid–liquid interface tension. A schematic of Wilhelmy plate method is shown in Fig. 5.9.

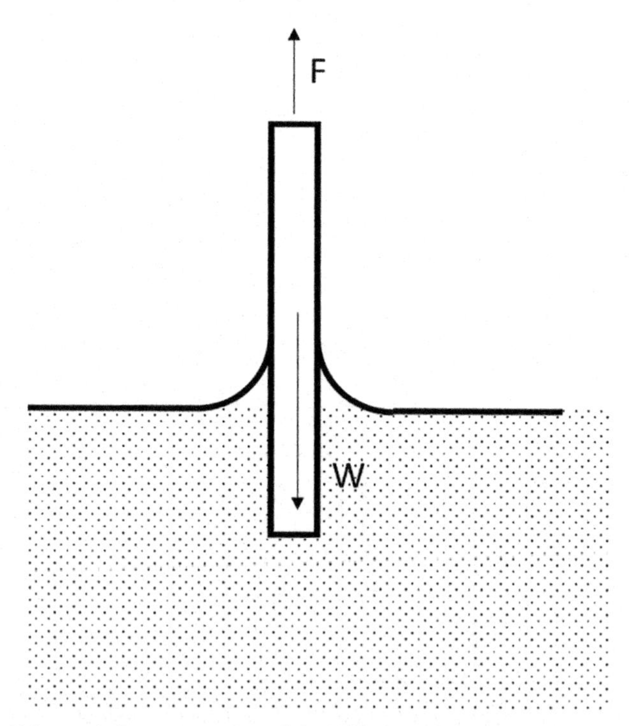

Figure 5.9 ***Schematic Representation of the Wilhelmy Plate Method.***

A tensiometer is normally used to measure the force, F required to pull the plate until it detaches from the water surface. When the plate's surfaces are completely wet, the equation below can be used to calculate the surface tension with accuracy of 0.1%.

$$\gamma = \frac{F - W}{2L}$$

A full description of the operating procedure can be found in ASTM D1331-14 Method C. According to the standard, at least two measurements should be recorded. More measurement should be recorded when variation in the surface tension measurements is significant.

ASTM D1331-14 Standard test methods for surface and interfacial tension of solution of paints, solvents, solutions of surface-active agents, and related materials-method C surface tension by Wilhelmy plate.

This test method provides information about surface tension determination of liquid materials which also include resin solutions. It requires no buoyancy correction and suitable for moderate viscosities (1 to 10 Pa s) solution.

2.1.3 Curing Characteristics

The chemical composition of liquid resin exists in the form of monomers consist of double bond. During curing, the resin monomers undergo polymerization process and the double bond disappears to form stable cross-linked bond. In the liquid-based system, UV radiation is normally used to initiate the photochemical reaction that enables the curing of resin. The process involves the conversion of liquid resin into cured resin and this process (speed and depth of curing) is affected by many factors. The factors include intensity and spectral range of illumination, thickness of curing area, distance between the light source and substrate, transparency of the substrate, and etc.

Several methods have been used to investigate the curing characteristics of liquid resin. These methods are:

- Differential scanning calorimetry (DSC).
- Dynamic mechanical analysis (DMA).
- Fourier transforms infrared spectroscopy (FT-IR).

DSC measures the heat produced by the photo-curing reaction by which the degree of cure can be determined as the reaction takes place. DSC is used to get data on the heat involved in various thermal processes, that is, its endo or exothermicity. The DSC measurements were performed dynamically using a liner temperature ramp. The samples are heated and cooled at a constant rate (10 K/min) and the different important parameters of the samples are measured as a function of temperature. Two heating runs are normally conducted. The measurements in the first run are used as baseline in quality control to remove the thermal history of the sample caused by processing conditions. The thermal history from the second run can be removed using the data obtained in the first run. DSC can measure glass transition temperature (T_g) even with smaller quantity of samples, crystallization peak temperature (T_c), melting peak temperature (T_m), melting enthalpy (D_{Hm}), crystallization enthalpy (D_{Hc}), degree of crystallinity on heating (X_m), and degree of crystallinity on cooling (X_c).

FT-IR provides more chemical information than DSC because many absorption bands are measured. DMA is a technique used to determine material property by giving small deformation to the sample in a cyclic manner [9]. It measures the modulus of the resin so that the properties of the cured material can be related to the degree of cure. An oscillatory force of known frequency is applied to a sample and the resultant deformation and phase lag recorded to obtain the modulus information. Normally, the sample can be subjected to constant stress or constant strain. For liquid resin, DMA can be done using a supporting structure, such as fiberglass braid. As sinusoidal force is applied, the obtained information can be expressed as an in-phase component, the storage modulus, and out-of-phase component, loss modulus. The calculated modulus can be used to deduce the degree of cure of the resin. A general instruction of the test method can be found in ASTM D4473-12.

ASTM D4473-12 Standard test method for plastics: dynamic mechanical properties: cure behavior.

This method covers the determination and reporting of the thermal advancement of cure behavior of thermosetting resin by using dynamic-mechanical-oscillation instrumentation. It provides a means for determining the cure behavior of thermosetting resins over a range of temperatures by free vibration as well as resonant and nonresonant forced-vibration techniques, in accordance with Practice D4065.

Fourier Transform Infrared (FT-IR) spectroscopy is used to obtain the infrared spectrum of absorption band of a substance. In FT-IR, the curing characteristic of the resin can be determined by monitoring the change in the absorption spectrum of the infrared radiation by the chemical bonds in the resin monomers. In typical UV-curable bond formation, the absorption spectrum of resin shows diminishing peaks around at certain wavenumbers as more double bonds is converted into single bond. Each peak is a result of the absorption by the resin monomer and the wavenumber of each peak corresponds to one of the bond vibration mode of the molecules. By monitoring the relative peak area of the absorption spectra, the curing kinematic and process and the reaction intermediates can be determined.

2.1.4 Thermal Stability

Aging is an important aspect of the liquid resin material. As the resin ages, there will be change in rheological property and chemical composition in the resin. For rheological property, it is found that the viscosity of the resin will increase with time. The increase in viscosity is a result of the physical and chemical aging of the resin. Hence, one crude and simple way to assess thermal stability of resin is by monitoring the viscosity change in the resin. The viscosity measurement techniques have been discussed previously. Otherwise, the physiochemical changes can be determined by using FT-IR and inverse gas chromatography (IGC) [10]. In IGC, the probe is injected into the column at constant carrier gas flow rate. The retention time of the resin vapor is then measured by gas chromatography detectors. By knowing the retention time, information, such as probe molecule size and concentration can be used to deduce the physiochemical properties of the resin. IGC is a nondestructive method and only assess the property at resin surface. Hence, FTIR is normally used for more sensitive and accurate result. The detection of chemical changes of resin by FTIR has been discussed previously. However, it is important to note that FTIR is not a quantitative method hence the result should be analyzed carefully.

2.1.5 Liquid Appearance

Liquid resin comes in forms of different appearances commonly described as clear, transparent, translucent, or in different colors as shown in Table 5.1.

Appearance assessment is a simple and easy way to estimate the quality of the liquid resin. A change in color indicates a change in wavelength absorption of visible light by the resin. Hence, it could be important to assess the appearance of the resin to ensure the quality of the resin as the change can indicate contamination or impurities in the raw material, process variations caused by heating and oxidation, or degradation of products exposed to weathering over time. There is no existing standard test method specifically for liquid resin. However, visual tests have been established by

Table 5.1 Examples of appearance of liquid resins

Techniques	Materials	Appearance	References
SLA	Somos 9920	Transparent Amber	[11]
Polyjet (inkjet printing)	RGD720	Transparent	[12]
	RGD837	White	
	RGD875	Black	
	RGD836	Yellow	
	RGD841	Cyan	
	RGD851	Magenta	

comparing the sample to a known standard. Objective measurements can also be done with color spectrophotometers that give reliable data on a consistent basis. ASTM 1544 and ASTM 5386 provide some information about the standard method to assess the color of any liquid materials.

D1544-04 Standard test method for color of transparent liquids (Gardner color scale).

This standard provides a standard test method for assessing the color of transparent liquid by comparing with arbitrarily numbered glass standards. This technique is a simple method as it requires only human judgment and do not require any sophisticated equipment. The standard also discussed about the calibration of the glass standards.

D5386-10 Standard test method for color of liquids using tristimulus colorimetry.

This standard covers instrumentation method for tristimulus colorimeter for the determination of color of clear liquid. It provides information about the required apparatus, standard procedures and calibration of the instrument. The measurement is converted to color ratings in the platinum cobalt system.

2.1.6 Liquid Density

The density of the liquid resin can be determined by following ASTM D3505. First, the resin sample is drawn into a calibrated bicapillary pycnometer of known weight. The weight of the resin is then measured when it reaches a reference temperature of 20°C. Consequently; the density of the resin is then calculated from the resin weight.

D3505-12 Standard test method for density or relative density of pure liquid chemicals.

This standard describes a simplified method for the measurement of density and relative density of pure liquid chemicals. It is normally used when thermal expansion function of the material is known however, it is also applicable to those which thermal expansion functions are not known. It provides information about the apparatus used, standard procedures, and calibration of the equipment and calculation of density.

Safety considerations for materials include the production of raw materials that should be nonhazardous. Robust methods and inspection techniques will be needed to certify the performance of raw materials for AM.

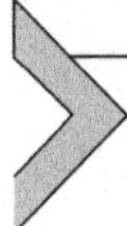

3 SOLID MATERIALS CHARACTERIZATION TECHNIQUES

3.1 Filament Diameter Consistency

FDM feedstock exists in the form of filament. The most commonly found FDM filaments exist have an average diameter of 1.78 mm, although diameter as large as 3 mm can still be found. Ideally, the filaments should have uniform diameter across the length. However, variance in the diameter is inevitable in the actual manufacturing process. Vogh et al. stated that both the average diameter and its variance are important and would affect the extrusion process [13]. The nonuniformity in diameter would lead to the improper feeding of feedstock and even nozzle clogging. Hence, it is important to ensure the consistency of the filament's diameter. The measurement of the diameter of filaments can be obtained using Vernier calipers, micrometer screw gage or any other measuring devices that have accuracies better than ± 0.005 mm.

3.2 Density

Density is defined as mass per unit volume. The density of thermoplastic filaments can be determined by using two techniques:

- Archimedes principle and
- density-gradient technique

To use Archimedes principle to measure density, the samples are weighed in air, before being immersed into a liquid with known density. The same samples are then weighed again, in the liquid. The density of the plastic sample can be calculated as:

$$\rho = \frac{A}{B} \times \partial$$

where ρ is the density of the solid, A is the weight of solid in air, B is the weight of solid in the liquid and ∂ is the density of the liquid. It is recommended that the experiment to be conducted at 23 ± 2°C and 50 ± 10 % relative humidity. A correction table is available to correct the density of the test liquid if the experiment is not conducted at 23°C. The details of this technique is documented in ASTM D 792.

ASTM D 792 Standard test methods for density and specific gravity (relative density) of plastics by displacement.

The standard covers the methods used to determine the specific gravity and density of solid plastics.

The density-gradient test method is designed to yield results with accuracy better than 0.05 %. This method requires a gradient tube preparation. A sample with unknown density and two sample floats with known densities (A and B) bracketing the sample's density. The height of the floats (y and z) and sample (x) from an arbitrary level are measured using a line through their center of volume. The density of the sample can be calculated using:

$$\rho = A + \left[\frac{(x-y)(B-A)}{(z-y)} \right]$$

The details of this method are documented in ASTM D1505.

ASTM D1505 Standard test method for density of plastics by the density-gradient technique.

This standard outlines the determination of the density of solid plastics by inspecting the position to which a test specimen submerges in a liquid column with varying gradient with the help of references of known density.

3.3 Porosity

Porosity is the void content of the filament measured in percentage. The void content of a filament may significantly affect some of its mechanical properties. Higher void contents usually cause lower fatigue resistance, more vulnerability to water penetration and weathering, and greater deviation in strength properties. The information on void content can be used as an estimation of quality of filament. Porosity of the filament can be determined via two methods and they are:

- fractography, and
- matrix removal method.

In fractography, a cross section of the filament is made smooth and then observed under an optical microscope [14,15]. Porosity is calculated as a ratio of the area of the void to the area of the material in the cross-section area of the filament.

Matrix removal method is normally used for determining the void content of composite filaments. ASTM 3171 has detailed the procedure for conducting the test.

ASTM 3171 Standard test methods for constituent content of composite materials.

The test method determines the constituent content of composite materials by digestion or ignition. It also allows the calculation of void volume in percentage.

Test method I of ASTM 3171 assumes that the reinforcement remains unchanged by the matrix removal process. There are several matrix removal processes, namely, by dissolution and combustion. In the dissolution process, the matrix material is dissolved leaving behind the residue, which is the reinforcement. The reinforcement is then filtered, washed, dried, cooled, and weighed. The weight percent of the reinforcement is calculated, and from this value, and if densities of both the composite and the reinforcement are known, the volume percent is calculated. An additional calculation for void volume may be made if the density of the matrix is known or determined.

$$v_v = 100 - (v_f + v_m)$$

where, V_v is the void volume, v_f is the fiber volume, and v_m is the matrix volume.

The acid digestion method is used by Chuang et al. to inspect the porosity of the as-received fiber filled filament [16]. Negative value of porosity was recoded and was likely due to the error in taking measurements.

3.4 Moisture Content

Moisture content has an impact on the melt viscosity, which may affect the print quality in FDM [17]. FDM thermoplastic filaments with low moisture content generally produce parts with better print quality [18] and with better mechanical properties [19]. The presence of water may also affect the bonding between the reinforcement material and the matrix material in the composite filament [20]. The equilibrium moisture content of the thermoplastic increases with increasing relative humidity, the moisture content of the thermoplastic can be controlled by conditioning it in a controlled environment with appropriate relative humidity. The quantitative determination of moisture content of thermoplastics can be obtained by loss in weight as described in ASTM D 6980.

ASTM D 6980 Standard test method for determination of moisture in plastics by loss in weight.

This test method allows the measurement of moisture by using reduction in weight technology down to 50 mg/kg. It can be used in most plastics.

Moisture analyzer with measuring sensitivity up to 0.0001 g and the capability of compensating lift caused by convection should be used for this test. 20–30 g of the test sample are first placed on the sample plan in the moisture analyzer and the test is run according to the guideline of the moisture analyzer. The test temperature is important and it is determined by conducting a single run of test with increasing temperature while other parameters remain unchanged. A graph of moisture content against temperature is plotted and the average temperature at which the moisture content remains low and constant is selected. This technique, however, might overestimate the moisture content especially for materials that are more volatile than water. A typical graph used for optimal test temperature selection is shown in Fig. 5.10.

3.5 Thermal Properties

Thermal properties such as thermal stability, melting point and glass transition point are important properties in FDM as FDM process involves

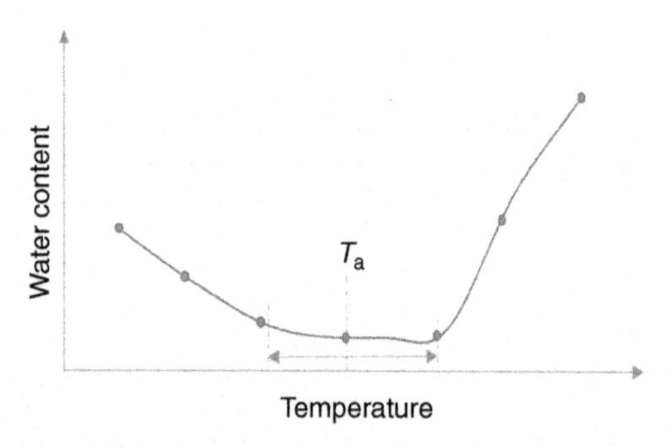

Figure 5.10 Optimal test temperature selection for quantitative determination of moisture content of thermoplastics (*arrow* indicates the range of temperature where the water content is low and constant, T_a is the average temperature that should be chosen for the test to determine the moisture content).

melting of the thermoplastics. It is well known that thermoplastics would degrade at high temperature. Hence, it is important to study the thermal stability of the thermoplastics at the working temperature of FDM, which is normally the melting point of thermoplastics. The techniques that are used in the study of thermal degradation include:

- Thermogravimetric analysis (TGA).
- Differential scanning calorimetry (DSC) [21].

TGA give information on the rate of degradation. In TGA, a sample of milligram size is heated, usually under an inert atmosphere, either isothermally, or at a constant rate, and the mass of sample is followed as a function of either time or temperature. The degradation relies on the rate at which the temperature is increased and this must be considered in any investigation. The emitted gases can also be analyzed by mass spectroscopy or infrared spectroscopy to give more information about the degradation. To have a more accurate determination of the temperature at which degradation occurs, derivative of the mass loss versus temperature is plotted in which the peak of the curve is when the rate of degradation is the highest. This method is called differential thermogravimetry (DTG).

A glass transition temperature (T_g) is useful in characterizing many important physical attributes of thermoplastic, thermosets, and semicrystalline materials including their thermal history, processing conditions, physical stability, the progress of chemical reactions, the degree of cure, and both mechanical and electrical behavior. There are two methods available to determine the glass transition temperature of the thermoplastics filament. The two methods include:

- Dynamic mechanical analysis (DMA).
- Differential scanning calorimetry (DSC).

The determination of Glass Transition Temperature (T_g) by dynamic mechanical analysis (DMA) according to ASTM E1640.

ASTM E1640 Standard test method for assignment of the glass transition temperature by dynamic mechanical analysis.

This method delineates the procedure to determine the glass transition temperature of materials having elastic modulus in the range of 0.5 MPa to 100 GPa using DMA.

A specimen of known geometry is placed in mechanical oscillation at either fixed or resonant frequency and changes in the viscoelastic response of the material are monitored as a function of temperature. Under ideal

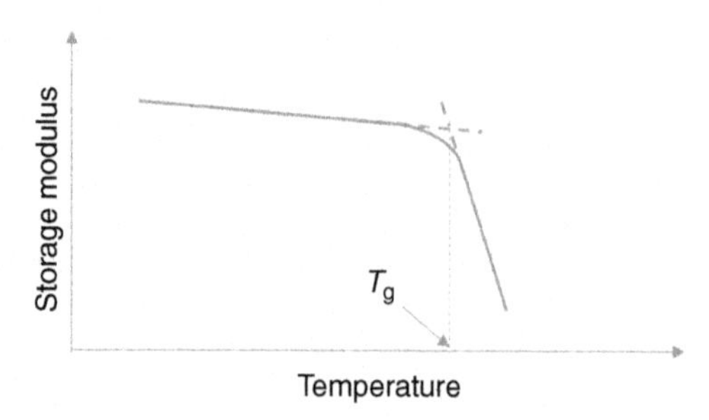

Figure 5.11 ***Determining Glass Transition Temperature in DMA.***

conditions, during heating, the glass transition region is marked by a rapid decrease in the storage modulus. The glass transition of the test specimen is indicated by the extrapolated onset of the decrease in storage modulus, which marks the transition from a glassy to a rubbery solid. A typical graph used to determine glass transition temperature in DMA is shown in Fig. 5.11.

In the study to formulate continuous carbon fiber reinforced polylactic acid composite, the dynamic storage modulus and loss tangent were analyzed to obtain the glass transition temperature [22]. The glass transition temperature is obtained by finding the corresponding temperature at the peak of the loss tangent curve.

The determination of glass transition temperature (T_g) by differential scanning calorimetry (DSC) is described in ASTM D3418.

ASTM D3418 Standard test method for transition temperatures and enthalpies of fusion and crystallization of polymers by differential scanning calorimetry.

This method is valid to polymers in any form from which appropriate specimens can be obtained.

A 10 mg test material is either heated or cooled at a controlled rate under a controlled flow of purge gas. The difference in heat input between a reference material and a test material due to energy changes in the material is recorded using suitable sensing device. A preliminary thermal cycle is performed to erase previous thermal history. It is performed by heating the test material at a rate of 20°C/min from 50°C under the melting temperature to 30°C above the melting temperature. To record the heating curve, the same test material is cooled to at least 50°C below the roughly estimated transition temperature. The heating is done at a rate of 20°C/min after holding the temperature for 5 minutes. The glass transition temperature

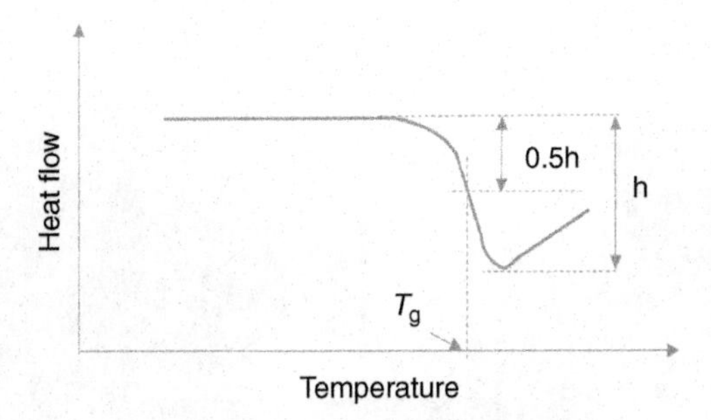

Figure 5.12 ***Determining Glass Transition Temperature in DSC.***

can be determined as the mid-point temperature between the extrapolated onset temperature and extrapolated end temperature. A typical graph used to determine the T_g using DSC is shown in Fig. 5.12.

TGA analysis and DSC were conducted by Francis and Jain to study the thermal characteristics of the polymer-layered silicate nanocomposite [23]. In another study, TGA was carried out to investigate the origin of blistering of Ultem 1000 filament [16]. TGA was able to determine water content in the Ultem resin pellet.

3.6 Microstructure of Composite Filament

Microstructure determines the mechanical properties of a material. Microstructure analysis is important especially for the composite filament in which interfacial properties between reinforcements and matrix material are of interest. Scanning electron microscopy is used to observe the interaction between the reinforcements and matrix material [8]. The sample is first coated with a thin layer of gold via plasma sputtering in order to eliminate charging effects on the surface and then secured on the stage in the chamber of the scanning electron microscope using a conductive tape. The chamber is then closed and vacuumed to low pressure. SEM images with resolution, as high as several nanometers can be taken.

Fig. 5.13 shows two SEM images of continuous carbon fiber reinforced thermoplastics taken in different magnifications. The interface between carbon fiber and the thermoplastic matrix can be clearly observed in Figs. 5.12 and 5.13B in which matrix and fibers of the filament can be found and there is also void at the center of the filament.

In the works of Gardner et al. printable carbon nanotube yarn filaments that consist of continuous fiber that infused with Ultem were fabricated

Figure 5.13 SEM images of markforged carbon fiber composite filament at (A) ×100 and (B) ×250 magnifications.

using continuous solution coating method [24]. The fabricated composite filament was put under the SEM to check the resin wet out of the inter bundle spaces. The uniformity of the coating can be observed using SEM. In another study, graphene composite made from graphite and ABS in *N*-methylpyrolidone was developed [25]. SEM images were taken to observe the incorporation of graphene sheets in the ABS matrix.

3.7 Mechanical Properties of Filament

The mechanical properties of the filament are important not only because they determine the mechanical properties of the printed parts, but they also affect the processability in FDM. Thermoplastics cannot be too brittle; otherwise the filament would break easily causing the printing process to fail. Mechanical properties such as the tensile strength and strain, torsional strength, and strain can be obtained.

In tensile test, specimens with the length of 380 mm were pulled in the tensile machine at constant crosshead speeds ranging from 0.00381 to 0.381 mm/s providing strain rates in the range of 10^{-5} to 10^{-3}/s. Tensile properties such as tensile modulus E, yield strength σ_{ys}, and strain ε_{ys}, values can be taken from the data. E is calculated as the gradient of the slope of the stress-strain curve, σ_{ys} is taken as the maximum stress reached during the test and ε_{ys} is the corresponding strain at maximum stress. Alternatively, string test can be used to the tensile properties of the FDM filament according to ASTM D638.

ASTM D638 Standard test method for tensile properties of plastics.

This test method contains the procedure to conduct tensile test on unreinforced and reinforced plastics in the form of standard dumbbell-shaped of any thickness up to 14 mm.

Figure 5.14 ***Schematic of Filament Tensile Test.***

It was found that the larger the diameter of the filament, the lower the tensile strength and modulus of the filament [26]. A schematic of filament tensile test is shown in Fig. 5.14.

In the torsion test, specimens with the length of 228 mm and uniform cross section are prepared [9]. The torsion test apparatus is setup as shown in Fig. 5.15.

Numerous masses were hung and the change in angle is measured with a protractor at the free end of the specimen. At least five readings for three specimens need to be taken for each mass value. G can be calculated via the relation after torque T and angular rotation $\Delta\theta$ data are collected during the tests:

$$\frac{T}{J} = G\frac{\Delta\theta}{L}$$

where, J is the polar moment of inertia of the cross-section area.

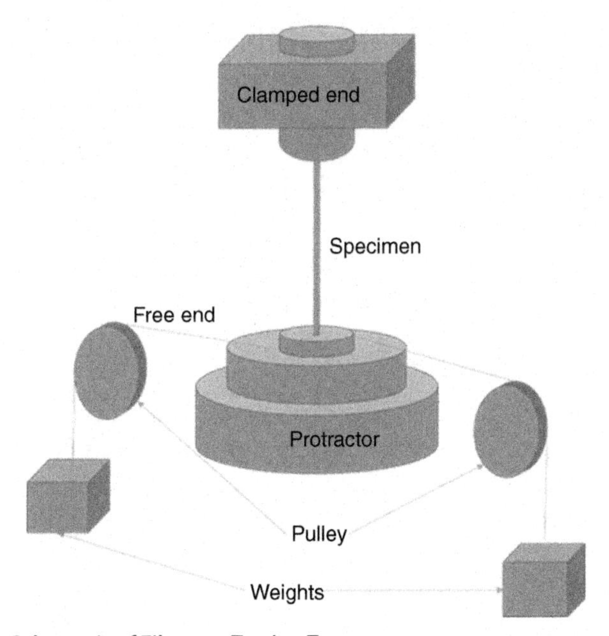

Figure 5.15 ***Schematic of Filament Torsion Test.***

3.8 Melt Flow Characteristics

Flow characteristics of the molten thermoplastics are very important in FDM technique. The flow characteristics are very much dependent on the viscosity of the molten thermoplastics. Thermoplastics that are too less viscous would have difficulty in forming the desired shape whereas thermoplastics that are too viscous would be hard to be extruded out of the nozzle. Melt flow index is used to determine the flow of the filament through the FDM machine [27–29]. Typically, melt flow index and viscosity are inversely related [30].

ASTM D1238 Standard test method for melt flow rates of thermoplastics by extrusion plastometer.

This method outlines the assignment of the rate of extrusion of molten thermoplastic resin using an extrusion plastometer.

Procedure A of ASTM D1238 can be used as a guideline to determine the melt flow rates of thermoplastics by extrusion plastometer. It is based on the measurement of the mass of material that extrudes from the loaded piston over a given period of time. The units of measure are grams of material/10 min (g/10 min). It is generally used for materials having melt flow rates that fall between 0.15 and 50 g/10 min. FDM can typically extrude materials with melt flow index around 2.41 g/10 min [31].

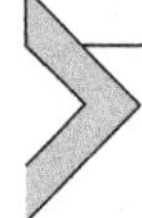

4 POWDER MATERIALS CHARACTERIZATION TECHNIQUES

The powders used for AM processes have certain requirements, which are critical to the success of the build process. The size distribution, morphology, chemical composition, flowability, density, and laser absorptivity of the powder particles are some of the most significant characteristics that influence the quality of the built part. In the following sub-sections, several methods will be introduced for each of the prior mentioned powder characteristics.

4.1 Powder Size Measurements

Powder particle sizes have a direct influence on the layer thickness and minimum feature size of an AM part. Smaller powder particles permit a thinner layer thickness, finer minimum feature size and better surface finish. After a sample was collected the powder size distribution can be determined using one of the several methods mentioned below:

- microscopy,
- sieving,

- gravitational sedimentation, and
- light scattering.

The first method of measuring powder size is to use microscopy techniques, such as optical microscopes, scanning electron microscopes and/or transmission electron microscopes. These techniques allow one to directly see and measure the various dimensions of the powder particles.

The second method is to use a series of sieves with different mesh openings to separate the powder into the different sizes.

ASTM B214-07 Standard test method for sieve analysis of metal powders.

This test method covers the dry sieve analysis of metal powders or mixed powders, using sieves with size between 45 and 1000 μm.

The sieves are stacked from top to bottom in the order of decreasing mesh opening and a collection pan is placed below the entire series of sieves. The shaker, which the entire set up is attached to, runs for 15 minutes and induces the sieving action.

The third particle size determination method is based on gravitational sedimentation

ASTM B761-06 Standard test method for particle size distribution of metal powders and related compounds by X-ray monitoring of gravity sedimentation.

This test method is applicable to particles of uniform density and composition, having particle size distribution between 0.1 and 100 m. The reported particle size measurement is a function of, both the actual particle dimension and shape factor, as well as the particular physical or chemical properties being measured.

In this method, the attenuation of a horizontally collimated beam of X-ray is measured after it passes through a liquid suspension containing powder particles of different sizes. Initially, the powder particles are homogenously mixed via circulation and the attenuation of the X-ray is at a maximum. Once circulation ceases, all particles begin settling to the bottom with the larger particles sinking at a faster rate than the others, as shown in Fig. 5.16. Under the laminar flow regime, the settling velocity of the particles can be directly related to an equivalent Stokes diameter through the Stokes equation. Therefore, particle sizes can be determined if the settling velocity of the particles can be obtained under a low Reynolds number flow. Since the vertical distance of the X-ray beam from the top of the cylinder is known, and the time taken for particles of different sizes to sink from the top of the cylinder to the X-ray beam can be obtained from the

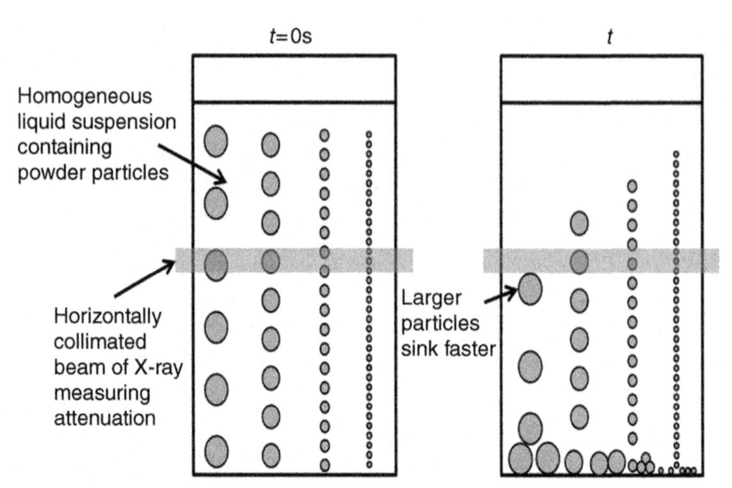

Figure 5.16 ***Illustration of Gravitational Sedimentation Method.***

X-ray attenuation plot, the particle velocities can be computed. Consequently, the particle size distribution can also be obtained.

The last method measures size using the information contained in the light scattered from the powder particles.

ASTM B822-10 Standard test method for particle size distribution of metal powders and related compounds by light scattering. This test method covers the determination of the particle size distribution by light scattering, reported as volume percent, of particulate materials including metals and compounds. It applies to analyse with both aqueous and nonaqueous dispersions. In addition, analysis can be performed with a gaseous dispersion for materials that are hygroscopic or react with a liquid carrier. The standard is applicable to the measurement of particulate materials in the range of 0.4–2000 μm, or a subset of that range, as applicable to the particle size distribution being measured.

Scattering can be achieved either by dispersing the particles in a liquid medium and circulating it through the light beam or by aspirating the dry particles in a carrier gas. The scattered light is captured by photodetector arrays, which then convert it to electrical signals for processing. By invoking the Mie Scattering or Fraunhofer Diffraction theory, or both, the collected signals can be converted into size distribution data.

4.2 Morphology

The morphology and shape of powder particles affect its flowability and the packing of the powder particles. Powder particles with poor

Table 5.2 Terms describing powder shapes in ASTM B243-11

Term	Definition
Acicular	Needle-shape
Flake	Flat or scale-like, whose thickness is small compared with other dimensions
Granular	Approximately equidimensional, nonspherical shapes
Irregular	Lacking symmetry
Needles	Elongated and rod-like
Nodular	Irregular, having knotted, rounded, or similar shaped
Platelet	Composed of flat particles having considerable thickness
Plates	Flat particles of metal powder having considerable thickness
Spherical	Globular-shaped

flowability may not spread evenly in powder fusion techniques such as selective laser sintering (SLS) and SLM. Irregularly shaped particles with low packing density could lead to low part density in the SLS process. Particle morphology can be evaluated using either a qualitative description of the particle shape or a quantitative number to represent a certain characteristic of the particle.

Under the qualitative powder morphology characterization method, a suitable description is given to the silhouettes of the observed particles. This is the simplest means to characterize the shape of the particles. In an effort to establish a common way of describing different particle shapes, a list of terms have been defined in ASTM B243-11 as shown in Table 5.2.

The quantitative method of characterization employs a single number to describe some certain features of the particle and it can be further classified into four categories: dimensional, sphericity, roundness, and perimeter. It should be noted that the disadvantage of a single-number classification is that particles with different morphology and shapes may end up with similar numbers for certain characteristics. Consequently, it is not possible to reconstruct the shape of the particle given its single-number value.

Under the dimensional category, the characteristics of large particles can be calculated using the length of the three orthogonal axes along with Eqs. 1 to 5 in Table 5.3.

On the other hand, only 2 dimensions are needed to adequately characterize the morphology of small particles. For instance, once the length and breadth of the particle is obtained such as using the silhouette of the particle, the length ratio and projected area ratio can be calculated using

Table 5.3 Table of formulae pertaining to single-number characterization of particle morphology

S/N	Equation	Description	Category	References
1.	$\frac{L+I}{2S}$	Flatness index	Dimensional	[32]
2.	$\frac{I}{L}, \frac{S}{I}$	Ordinate and Abscissa for a Plot to Characterize Shape	Dimensional	[32]
3.	$\frac{I.100}{L}$	Elongation	Dimensional	[32]
4.	$\frac{S.100}{L}$	Flatness	Dimensional	[32]
5.	$\frac{S}{L}$	Flatness	Dimensional	[32]
6.	$\frac{L}{B}$	Length ratio	Dimensional	[33]
7.	$\frac{\text{Projected area of silhouette}}{\text{area of rectangle BL}}$	Projected area ratio	Dimensional	[33]
8.	$\frac{\text{Diameter of a circle equal in area to that of the particle silhouette}}{\text{Diameter of the smallest circle circumscribing the particle silhouetter}}$	Sphericity	Sphericity	[34]
9.	$\frac{4\pi.\text{area of particle silhouette}}{(\text{perimeter of particle silhouette})^2}$	Sphericity	Sphericity	[35]
10.	$\frac{\sum_{i=1}^{N} r_i}{N} \cdot \frac{1}{R}$	Roundness	Roundness	[36]
11.	$\frac{\text{Radius of curvature of the most convex part}}{(L+B)/4}$	Roundness	Roundness	[37]
12.	$\frac{\text{Radius of curvature of the most convex part}}{L/2}$	Roundness	Roundness	[37]
13.	$\frac{(\text{perimeter of the particle outline})^2}{4\pi.(\text{area of particle silhouette})}$	Circularity shape factor	Perimeter	[37]

Eqs. 6 and 7 in Table 5.3 respectively. Sphericity, which is a measure of the particle's resemblance to a sphere, can be calculated easily using either Eqs. 8 or 9 in Table 5.3.

Roundness, which is a measure of the smoothness of the particle perimeter, can be calculated using Eq. 10 in Table 5.3. The radius of curvature (r), total number of protrusion along the particle silhouette (N) and the radius of the maximum inscribed circle (R) used in Eq. 10 in Table 5.3 are illustrated in Fig. 5.17.

Eqs. 11 and 12 in Table 5.3 can also be used to characterize roundness. The inputs to these equations, such as the greatest length (L) and breadth (B) of the particle are illustrated in Fig. 5.18.

The last parameter is the circularity of the particle and it can be calculated using a widely accepted Eq. 13 in Table 5.3. This parameter measures the particle's resemblance to a circular profile and is in essence similar to sphericity.

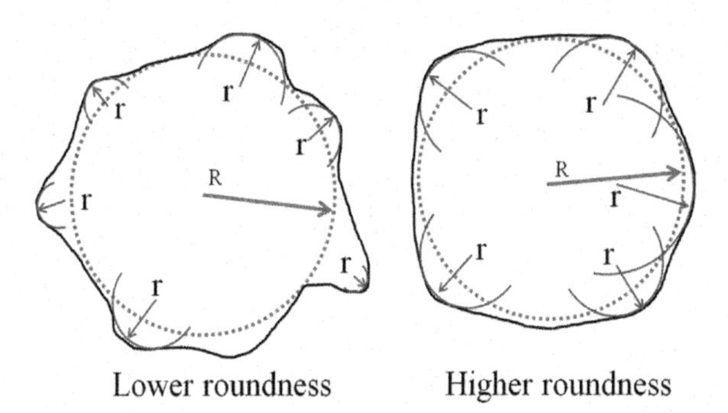

Figure 5.17 ***Illustration of Technique Used by Wadell to Obtain Roundness Values.***

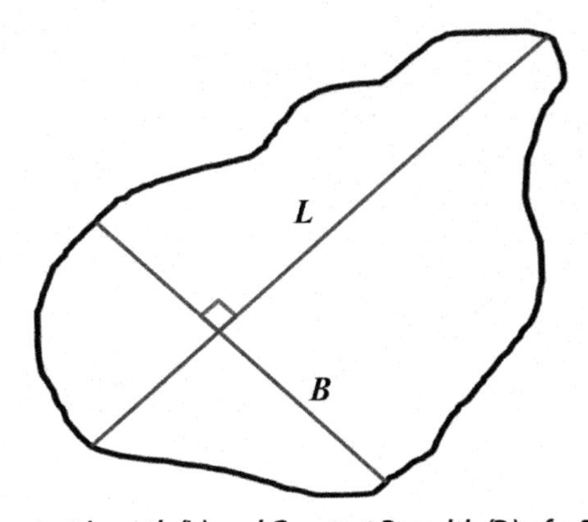

Figure 5.18 ***The Greatest Length (L) and Greatest Breadth (B) of a Particle Silhouette.***

4.3 Chemical Composition

The mechanical and functional properties of AM parts are primarily dependent on the material composition. Furthermore, quality of the powder used in powder bed fusion processes may also degrade with repeated usage and recycling. It is, therefore important to evaluate the composition of powder regularly. The methods for chemical composition analysis can be divided into:

- microanalysis,
- surface analysis, and
- bulk analysis.

Energy dispersive X-ray spectroscopy is a microanalysis method that uses a beam of X-ray to excite the electrons in the analyzed sample. The excitation can cause an electron to be either removed from an atom or promoted to a higher energy level. Either way, a hole is created in the energy level where the electron once was. This hole will eventually be filled by an electron from the higher energy level. As the electron from the outer shell drops to the hole in the inner shell, an X-ray with energy equal to the difference between the two energy levels is emitted and subsequently captured by the energy dispersive spectrometer. Since the energy of emitted X-ray is unique, different elements can be identified. Fig. 5.19 shows a plot of the X-ray intensity against energy with the respective elements labeled over the different energy peaks. The relative amount of each element in weight or atomic percentage is calculated using the intensities of the peaks. This elemental composition measurement can also be similarly performed using an electron beam instead of an X-ray beam.

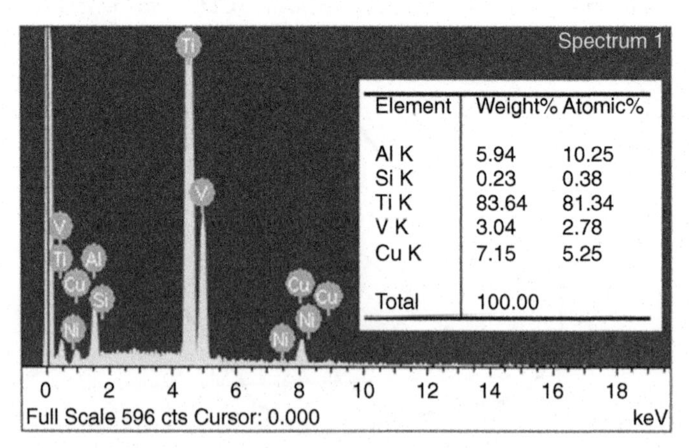

Element	Weight%	Atomic%
Al K	5.94	10.25
Si K	0.23	0.38
Ti K	83.64	81.34
V K	3.04	2.78
Cu K	7.15	5.25
Total	100.00	

Figure 5.19 ***Output From Energy Dispersive X-Ray Spectroscopy Measurements.***

Surface analysis methods consist of atomic emission spectroscopy, X-ray photoelectron spectroscopy and secondary ion mass spectrometry. In atomic emission spectroscopy, excitation of atoms is achieved by imparting energy in the form of a spark, flame, plasma, or arc. The excited atoms give off various wavelengths of light, which is collected and analyzed. Since the energies and wavelengths of the emitted light are unique to each element, identification can be carried out this way. The composition of the sample can be determined from the intensities of the peaks.

X-ray photoelectron spectroscopy uses a monochromatic incident beam of X-ray to eject electrons out of atoms on the surface of a sample. Subsequently, the kinetic energies of the photoelectrons are measured and used to calculate its binding energy. By the conservation of energy, the energy of the incident X-ray necessarily equals to the sum of the photoelectron's kinetic energy and binding energy. Once the binding energy is known, it can be matched to a specific energy level of the corresponding element. Like all the previous methods, the binding energy determines the element while the intensity determines the composition.

In secondary ion mass spectrometry (SIMS), a primary ion beam is incident on the sample surface to eject secondary ions, which will be used for analysis. The mass of the secondary ions is obtained via a mass analyzer and subsequently used to identify the elements, isotopes or molecules present on the sample surface. Fig. 5.20 shows a simple illustration of the working principle of SIMS. This technique can detect all elements in the periodic table with a high sensitivity ranging from parts per million to parts per billion.

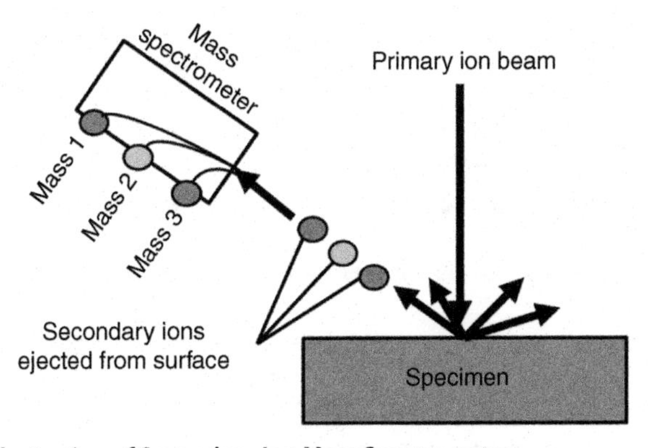

Figure 5.20 ***Illustration of Secondary Ion Mass Spectrometry.***

The bulk analysis methods include X-ray diffraction, X-ray fluorescence, inert gas fusion, and atomic absorption spectroscopy, inductively coupled plasma optical emission spectroscopy. In X-ray diffraction, the incident X-ray reflects off planes of atoms in the sample and interacts with each other resulting in interference patterns. Whenever the optical path difference between two interacting X-rays is equal to an integer number of its wavelength, constructive interference occurs and an intense reflection is detected at a particular grazing angle. Since the wavelength of the X-ray and the grazing angle are known, Bragg's law can then be used to solve for the perpendicular distances between different planes of atoms within a crystalline material. Once the positions of the different planes are known, the atoms can then be located at the intersection points of the planes. This crystal lattice information allows for phase identification. A schematic of X-ray diffraction on atoms is shown in Fig. 5.21.

X-ray fluorescence is identical to energy dispersive X-ray spectroscopy where a primary X-ray beam is incident onto the sample surface to emit secondary X-rays. The energy of the secondary X-ray is characteristic of the element while the intensity indicates the elemental concentration.

Inert gas fusion is a technique that is used to quantify the amount of oxygen, nitrogen and hydrogen content in a metallic sample. The procedure is as follows: firstly, a graphite crucible is heated to 3000°C within an inert gas flow for the purpose of decontamination. Secondly, the temperature of the crucible is lowered. Thirdly, the sample is lowered into the hot crucible and melted. Fourthly, the nitrogen and hydrogen elements in the molten

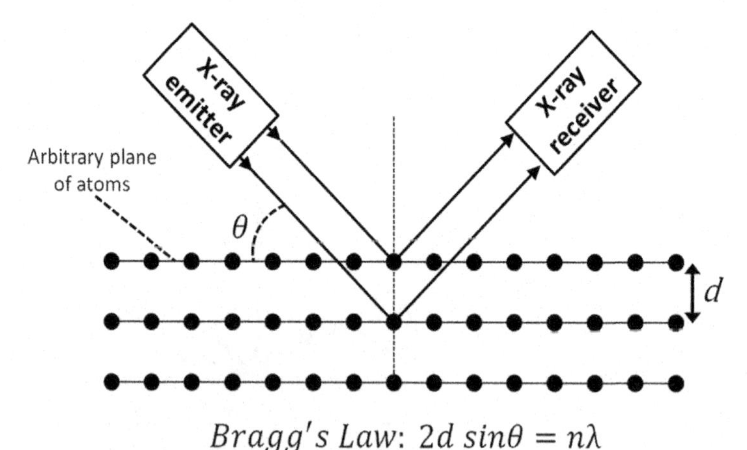

Figure 5.21 ***Illustration of X-Ray Reflecting off Planes of Atoms.***

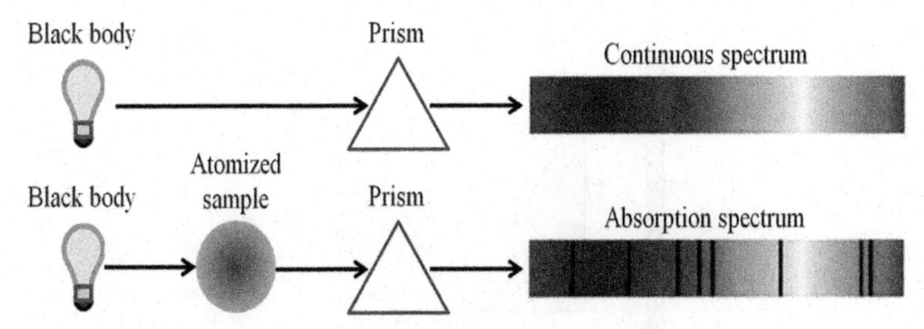

Figure 5.22 ***Illustration of Atomic Absorption Spectroscopy.***

sample get released as gas while the oxygen atoms react with the graphite crucible to form either CO or CO_2. Lastly, a detector determines the amount of CO, CO_2, N_2, and H_2 within the inert gas flow to evaluate the original concentrations of those elements in the sample.

Atomic absorption spectroscopy relies on the principle that electrons in an atom absorb and emit light of specific wavelength that is characteristic of the element. When a continuous spectrum of radiation is passed through the atomized sample, certain wavelengths of light will be absorbed and lost. This results in a pattern of dark lines in the absorption spectrum, which can then be used for the identification of elements. By comparing the difference in the detected radiation flux with and without the atomized sample, the elemental concentration of the sample can be determined using the Beer–Lambert law. A schematic of atomic absorption spectrometry is shown in Fig. 5.22.

In inductively coupled plasma optical emission spectroscopy, atomic excitation is achieved by introducing the liquid sample into plasma at 10,000 K. The excited atoms or ions subsequently relax to lower energy states by emitting photons with energies that are characteristic of the element. Determinations of the photon energies allow the elements to be identified while the intensities of the various detections give information about the concentrations of each element.

4.4 Flow Characteristics

The flowability of the powder is an important characteristic for powder bed fusion techniques. It affects the uniformity of the recoated powder in every layer. Poor flowability is likely to be caused by high interparticle frictional forces, which could in turn lead to low packing density, as well as part density. There are 2 types of flowmeter, which can be used to determine the powder flow rate.

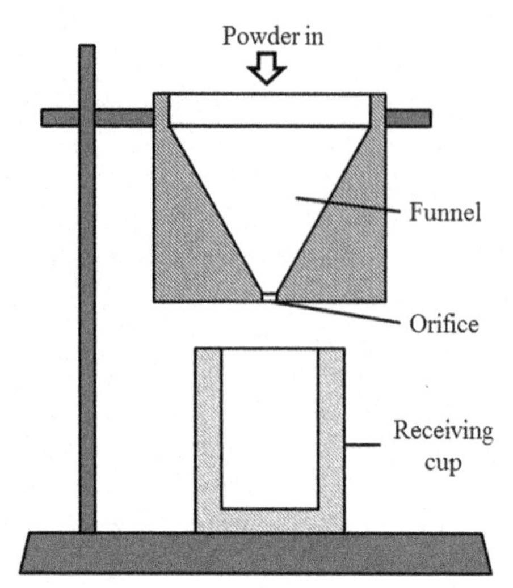

Figure 5.23 ***Schematic of Hall and Carney Flowmeter Design.***

ASTM B213-11 Standard test methods for flow rate of metal powders using the Hall flowmeter funnel.

This test method covers the determination of the flow rate of metal powders and is suitable only for those powders that will flow unaided through the specified apparatus, which has an orifice of 2.54 mm.

ASTM B964-09 Standard test methods for flow rate of metal powders using the carney funnel.

These test methods cover the determination of a flow rate, by use of the Carney funnel, which has an orifice of 5.08 mm, of metal powders and powder mixtures that do not readily flow through the Hall funnel.

The standard funnel design is illustrated in Fig. 5.23.

Powder flow can be measured using either a static or dynamic flow method. In the earlier case, a dry finger is used to cover the orifice while the powder is poured into the flowmeter. The measurement time starts as soon as the finger is removed and stops when all the powder has flowed through. Under the dynamic flow method, the orifice is left open during powder loading and the timing starts when the powder is first seen exiting the orifice and stops when all the powder has flowed through. The mass flow rate is determined by dividing the flow time in seconds by the powder mass in grams. Volumetric flow rate can be calculated in a similar manner by dividing the flow time by the volume of powder. However, the volume of the powder will need to be measured using an Arnold meter or any of the methods described in the density measurement section.

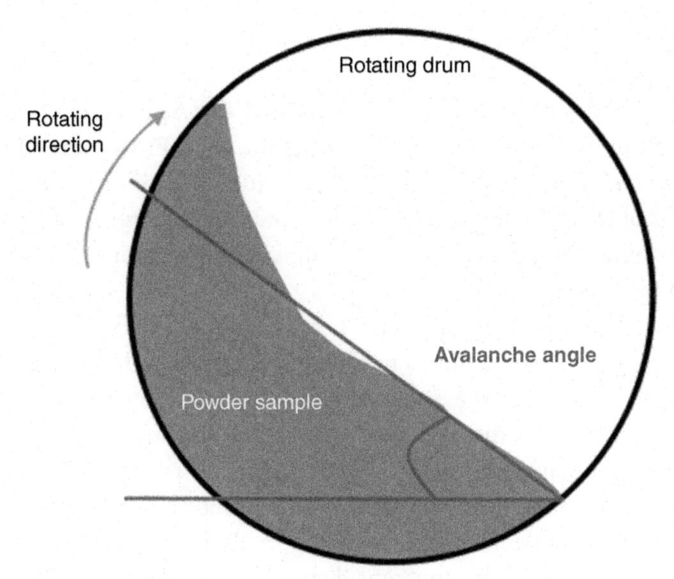

Figure 5.24 ***Schematic of Powder Avalanche Angle.***

In addition to the flowmeters, a revolution powder analyzer can also be used to indicate flowability by measuring the avalanche angle of the powder. Lower avalanche angles are indicative of better flowability of powders [38]. In this test the powder is placed inside a cylindrical drum with transparent glass sides. The drum was then set to rotate and a digital camera is used to monitor the flow behavior of the powder. Due to the drum rotation, the powder would be carried up along the side of the drum until it could not support its weight, forming avalanches. The avalanche angle is then computed by measuring the angle where the powder was at maximum position before the start of the avalanche. A schematic of avalanche angle is shown in Fig. 5.24.

4.5 Density

The packing density of powders could have a significant effect on the eventual bulk density of AM parts. There are three different types of densities that can be measured for powders:

- apparent density,
- tap density, and
- skeletal density.

The apparent density is an important measure of a material characteristic of the powder that is useful to the powder producers and powder users in determining quality and lot-to-lot consistency. The apparent density can be measured according to ASTM B212-09 using the Hall flowmeter, which is also used for flow rate measurements.

ASTM B212-09 Standard Test Method for Apparent Density of Free-Flowing Metal Powders Using the Hall Flowmeter Funnel. This test method describes a procedure for determining the apparent density of free-flowing metal powders and mixed powders. However, it is only suitable for powders that will flow unaided through the specified Hall Flowmeter funnel.

The powder that is loaded into the funnel is allowed to flow into a collection cup of known volume until a mound is formed. After that a nonmagnetic spatula is used to remove the excess powder such that the remaining powder flushes with the top of the cup. The mass of the powder in the cup is then measured and the apparent density is calculated by taking the ratio of the measured powder mass to the known volume of the collection cup.

The Scott volumeter, as shown in Fig. 5.25, is another apparatus that can be used to measure the apparent density of powders. Compared to the

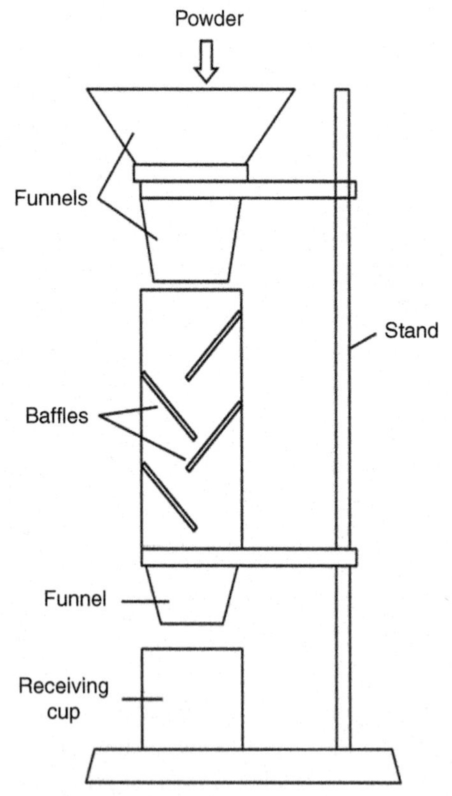

Figure 5.25 ***Illustration of Scott Volumeter.***

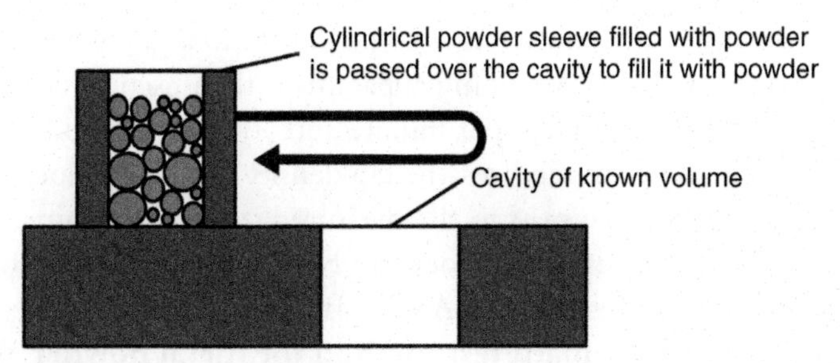

Figure 5.26 ***Illustration of Arnold Meter.***

Hall flowmeter, it has additional series of funnels and baffles to ensure that the loaded powder detaches into loose particles before entering the collection cup. The apparent density measurement and calculation procedures are exactly identical as those described in the previous paragraph when a Hall flowmeter is used.

The apparent density can also be measured using an Arnold meter, as shown in Fig. 5.26, is described in ASTM B703-10.

ASTM B703-10 Standard test method for apparent density of metal powders and related compounds using the Arnold meter.

This test method is applicable to free-flowing and nonfree-flowing metal powders, lubricated powder mixtures, and metal compounds. This test method simulates the action of the feed shoe on a powder compacting press and gives an apparent density value that closely approximates the apparent density of the powder in the die cavity after the production filling operation.

The Arnold meter is simply a steel block with a cavity of known volume located in the middle. In addition, it has a sleeve, which is used for powder delivery. Powder is first loaded into the sleeve, which is placed next to the cavity. The sleeve is then slid across the cavity to fill it up with powder up to the level where it is flushed with the upper surface of the steel block. The mass of the powder in the cavity is then measured and the apparent density is calculated by taking the ratio of the powder mass to the known volume of the cavity.

The procedure to measure tap density is described in ASTM B527-06.

ASTM B527-06 Standard test method for determination of tap density of metallic powders and compounds.

This test method specifies a method for the determination of tap density (packed density) of metallic powders and compounds, that is, the density of a powder that has been tapped, to settle contents, in a container under specified conditions.

It starts by pouring a known mass of powder into a graduated cylinder with volumetric markings. A tapping apparatus is then used to tap the cylinder at a rate between 100 taps per minute and 300 taps per minute. Once the volume of the powder stabilizes, the tap density can be obtained by taking the ratio of the known mass to the measured stabilized volume.

The skeletal density of the powder can be measured using nitrogen or helium pycnometry as described in ASTM B923-10.

ASTM B923-10 Standard test method for metal powder skeletal density by helium or nitrogen pycnometry.

This test method covers determination of skeletal density of metal powders. The test method specifies general procedures that are applicable to many commercial pycnometry instruments. The method provides specific sample outgassing procedures for listed materials. It includes additional general outgassing instructions for other metals. The ideal gas law forms the basis for all calculations.

A powder of known mass is first put into a sample chamber of known volume that is subsequently evacuated to a state of vacuum. Next, a container housing helium or nitrogen at a known pressure and volume is released into the sample chamber such that the inert gas permeates the gaps between the powder particles. Using the Ideal Gas Law together with the known initial and final states of the entire system, the total volume of the powder particles (excluding the gaps between particles) can be obtained. The ratio of the known mass to the obtained powder volume gives the skeletal density. The skeletal density should be similar to the bulk density of the material if there are no significant porosities embedded within the powder particles.

4.6 Laser Absorption Characteristics of Powder

The absorptivity of the powder (values between 0.0 and 1.0) particles for a laser of a particular wavelength is an important parameter. It directly influences the laser intensity setting needed to achieve a well-consolidated and dense part in powder bed fusion techniques. Powders with low absorptivity can only be fused at higher laser intensities and vice versa. Furthermore, the absorptivity of the powder is also a needed input for some computer simulations on powder bed fusion techniques.

The absorptivity of powder particles can be obtained through:

- computer calculations: ray tracing method and
- experimental techniques.

Ray tracing calculation accounts for the reflection of laser beam off the powder particle. The multiple reflections that occur within the powder

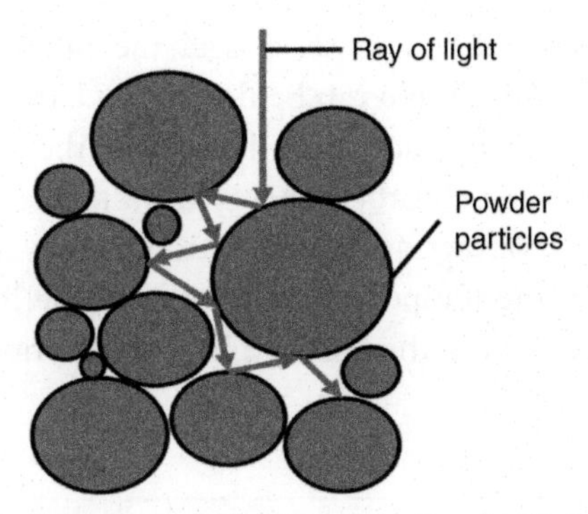

Figure 5.27 ***Multiple Reflections of Light Within a Powder Bed.***

bed results in a three-dimensional absorption of laser energy. The fraction of energy that is absorbed when a ray of laser beam incidents on a particle can be determined using the Fresnel equations. By combining the laws of reflection together with the Fresnel equations the overall absorptivity of the powder bed can be determined. The absorptivity of a powder bed is usually higher than that of a flat surface due to the multiple reflections and absorption within the powder bed, as shown in Fig. 5.27.

The experimental setup that can be used to measure the absorptivity of metallic powders under 1 μm wavelength laser source has been demonstrated by Wu et al. [39] and the simplified illustration of the experimental setup is shown in Fig. 5.28.

A uniform laser source is used to illuminate the powder layer resting on a thin piece of refractory metal base. The temperature rise of the powder sample and refractory metal base is captured by the thermocouples attached to the bottom of the base. Using the temperature readings along with the

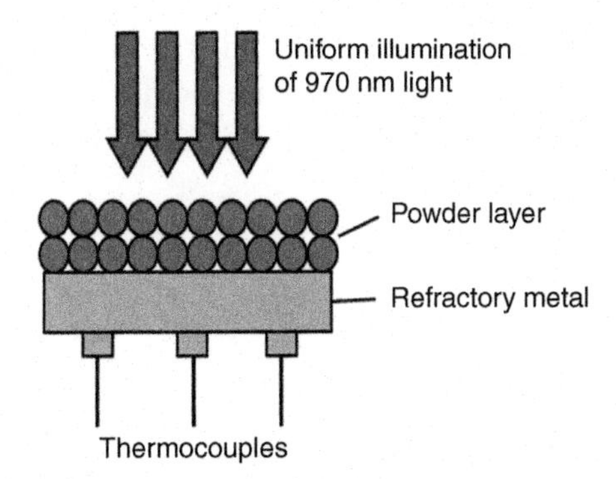

Figure 5.28 ***Illustration of Experimental Setup Used to Measure Powder absorptivity.***

powder and refractory metal base material data, the amount of energy that is being accumulated in the sample can be determined. By the conservation of energy, the amount of energy accumulating within the powder and metal base necessarily equal to the absorbed energy from the laser source less the thermal losses through convection and radiation. Using this conservation relationship along with the temperature data during the heating (laser on) and cooling (laser off) periods, the absorptivity of the powder samples can be obtained.

5 FUTURE OUTLOOK

There are readily available techniques and corresponding standards for the characterization of materials for AM, however, the suitability and adaptability of these specifically for AM remain to be evaluated. Materials characterization is important in AM as it provides insights to dependence of the part properties fabricated by AM on the raw materials properties, hence, there is a need to develop appropriate or new measurement techniques and standards.

Many AM processes are open source processes, in a sense that their process parameters can be varied and high-equality parts can be fabricated using a range of process parameters, instead of a specific parameters set. Predictive models can help to correlate raw material properties and processing parameters to the properties of the end parts. However, this will need wider collection of data on materials properties.

Critical factors such as longevity, durability and recyclability of AM materials should also be carefully considered and documented. For example, the re-usability of powder in powder bed fusion technology should be carefully evaluated, such as the ratio of virgin to reused powder and the effect of processing on the reused powder properties.

Safety considerations for materials include the production of raw materials that should be non-hazardous. Robust methods and inspection techniques will be needed to certify the performance of raw materials for AM.

6 QUESTIONS

1. State the different types of materials used as feedstock in additive manufacturing. For each type, suggest and describe an additive manufacturing technique.

2. Define the term "viscosity" and briefly explain the measurement techniques that can be used to characterize this property.
3. Explain Archimedes' Principle and state what can be quantified using this principle.
4. Suggest reasons why powder size and powder size distribution are critical for powder based additive manufacturing.
5. State the terms used to describe powder morphology and explain in details, what is "roundness" of a powder particle.
6. Explain the different techniques used in measuring density, clearly stating the difference between tap density, apparent density, and skeletal density.

REFERENCES

[1] K.R. Symon, Mechanics, Addison-Wesley Publishing Company, Boston, USA, (1971).
[2] A. Bastian. (2015). Continuous top-down DLP experiments. Available from: http://www.instructables.com/id/Continuous-Top-Down-DLP-Experiments/step2/Results-and-Conclusions/
[3] A.M. Elliott, The effects of quantum dot nanoparticles on polyjet direct 3D printing process, Doctoral Dissertation, Mechanical Engineering, Virginia Tech, Virginia Polytechnic Institute and State University, 2014.
[4] M. Vaseem, G. McKerricher, A. Shamim, 3D inkjet printed radio frequency inductors and capacitors, in: Microwave Integrated Circuits Conference (EuMIC), 2016 11th European, 2016, pp. 544–547.
[5] T. Rechtenwald, A. Kopczynska, E. Schmachtenberg, M. Devrient, T. Frick, and M. Schmidt, Investigation of material compatibility for embedding stereolithography, in: Proceeding of the 5th Multi-Material Micro Manufacture, Cardiff University, Cardiff, UK, 2008, p. 4.
[6] Schott Instrument. Capillary viscometry from SCHOTT instruments Available from: http://www.sartorom.ro/sites/default/files/produse/documente/73232_Lab-Products_No-6_Capillary-Viscometry_2-MB_English-pdf.pdf
[7] Spectro Scientific, Guide to measuring oil viscosity, in: Spectro Scientific, 2016.
[8] F. Ning, W. Cong, J. Qiu, J. Wei, S. Wang, Additive manufacturing of carbon fiber reinforced thermoplastic composites using fused deposition modeling, Compos. Part B: Eng. 80 (2015) 369–378.
[9] J.F. Rodríguez, J.P. Thomas, J.E. Renaud, Mechanical behavior of acrylonitrile butadiene styrene (ABS) fused deposition materials. Experimental investigation, Rapid Prototyp. J. 7 (2001) 148–158.
[10] B. Strzemiecka, B. Borek, A. Voelkel, Assessment of resin adhesives aging by means of rheological parameters, inverse gas chromatography, and FTIR, J. Adhesion Sci. Technol. 30 (2016) 56–74.
[11] Techok INC. Somos 9920 data sheet. Available from: http://www.techok.com/pdf/somos9920.pdf
[12] Stratasys. Digital material data sheet. Available from: http://usglobalimages.stratasys.com/Main/Files/Material_Spec_Sheets/MSS_PJ_DigitalMaterialsDataSheet.pdf?v=635796522191362278
[13] B. D. Vogt, F. Peng, E. Weinheimer, M. Cakmak, FDM from a polymer processing perspective: challenges and opportunities. Available from: https://www.nist.gov/sites/default/files/documents/mml/Session-4_2-Vogt.pdf

[14] D. Fitz-Gerald, J. Boothe, Manufacturing and characterization of poly (lactic acid)/carbon black conductive composites for FDM feedstock: an exploratary study, Cali Poly (2016) Student research, senior project, materials engineering, 152, 1-22.
[15] M.L. Shofner, K. Lozano, F.J. Rodríguez-Macías, E.V. Barrera, Nanofiber-reinforced polymers prepared by fused deposition modeling, J. Appl. Polym. Sci. 89 (2003) 3081–3090.
[16] K.C. Chuang, J.E. Grady, S.M. Arnold, R.D. Draper, E. Shin, C. Patterson, et al., A fully nonmetallic gas turbine engine enabled by additive manufacturing, part II: Additive manufacturing and characterization of polymer composites, NASA Glenn Research Center; Cleveland, OH United States, Technical Report, 2015.
[17] J.L. Willett, B.K. Jasberg, C.L. Swanson, Rheology of thermoplastic starch: effects of temperature, moisture content, and additives on melt viscosity, Polym. Eng. Sci. 35 (1995) 202–210.
[18] S.N. A.M. Halidi J. Abdullah, Moisture effects on the ABS used for fused deposition modeling rapid prototyping machine, in: 2012 IEEE Symposium on Humanities, Science and Engineering Research, 2012, 839–843.
[19] E. Kim, Y.J. Shin, S.H. Ahn, The effects of moisture and temperature on the mechanical properties of additive manufacturing components: fused deposition modeling, Rapid Prototyp. J. 22 (2016) 887–894.
[20] G.H. Yew, A.M.M. Yusof, Z.A.M. Ishak, U.S. Ishiaku, Water absorption and enzymatic degradation of poly (lactic acid)/rice starch composites, Polym. Degrad. Stability 90 (2005) 488–500.
[21] C. Wilkie, M.A. McKinney, Thermal properties of thermoplastics, in: J. Troitzsch (Ed.), Plastics Flammability Handbook: Principles, Regulations, Testing, and Approval, Hanser, Munich, 2004, pp. 58–76.
[22] N. Li, Y. Li, S. Liu, Rapid prototyping of continuous carbon fiber reinforced polylactic acid composites by 3D printing, J. Mater. Process. Technol. 238 (2016) 218–225.
[23] V. Francis, P.K. Jain, Experimental investigations on fused deposition modelling of polymer-layered silicate nanocomposite, Virtual Phys.l Prototyp. 11 (2016) 1–13.
[24] J.M. Gardner, G. Sauti, J.-W. Kim, R.J. Cano, R.A. Wincheski, C.J. Stelter, et al. 3-D printing of multifunctional carbon nanotube yarn reinforced components, Add. Manufact. 12 (2016) 38–44 Part A.
[25] X. Wei, D. Li, W. Jiang, Z. Gu, X. Wang, Z. Zhang, et al. 3D Printable Graphene Composite, Sci. Rep. 5 (11181) (2015) 1–7.
[26] J. N. Baucom, A. Rohatgi, W.R. Pogue III, and J. P. Thomas, Characterization of a multifunctional liquid crystalline polymer nanocomposite, presented at the 2005 SEM Annual Conference & Exposition on Experimental and Applied Mechanics, Portland, OR, 2005.
[27] T.N.A.T. Rahim, A.M. Abdullah, H.M. Akil, D. Mohamad, Z.A. Rajion, Preparation and characterization of a newly developed polyamide composite utilising an affordable 3D printer, J. Reinforced Plastics Comp. 34 (2015) 1628–1638.
[28] H. Garg, R. Singh, Investigations for melt flow index of Nylon6-Fe composite based hybrid FDM filament, Rapid Prototyp. J. 22 (2016) 338–343.
[29] N.M.A. Isa, N. Sa'ude, M. Ibrahim, S.M. Hamid, K. Kamarudin, A study on melt flow index on copper-abs for fused deposition modeling (FDM) feedstock, Appl. Mech. Mater. 773–774 (2014) 8–12.
[30] F. Cruz, S. Lanza, H. Boudaoud, S. Hoppe, M. Camargo, Polymer recycling and additive manufacturing in an open source context: optimization of processes and methods, in: Annual International Solid Freeform Fabrication Symposium, Austin, Texas, 2015, pp. 1591–1600.
[31] R. Singh, P. Bedi, F. Fraternali, I.P.S. Ahuja, Effect of single particle size, double particle size and triple particle size Al_2O_3 in Nylon-6 matrix on mechanical properties of feed stock filament for FDM, Comp. Part B: Eng. 106 (2016) 20–27.

[32] P.J. Barrett, The shape of rock particles, a critical review, Sedimentology 27 (1980) 291–303.
[33] Q. Guo, X. Chen, H. Liu, Experimental research on shape and size distribution of biomass particle, Fuel 94 (2012) 551–555.
[34] J.W. Bullard, E.J. Garboczi, Defining shaper measures for 3D star-shaped particles: sphericity, roundness, and dimensions, Powder Technol. 249 (2013) 241–252.
[35] P. Vangla, G.M. Latha, Influence of particle size on the friction and interfacial shear strength of sands of similar morphology, Int. J. Geosynth. Ground Eng. 1 (2015).
[36] R.D. Hryciw, J. Zheng, K. Shetler, Particle roundness and sphericity from images of assemblies by chart estimates and computer methods, J. Geotechn. Geoenviron. Eng. 142 (2016).
[37] A.E. Hawkins, , The shape of powder-particle outlinesvol. 1Research Studies Press Ltd, (1993).
[38] S.L. Sing, W.Y. Yeong, F.E. Wiria, Selective laser melting of titanium alloy with 50 wt% tantalum: Microstructure and mechanical properties, J. Alloys Compounds 660 (2016) 461–470.
[39] S. Wu, I. Golosker, M. LeBlanc, S. Mitchell, A. Rubenchik, J. Stanley, et al., Direct Absorptivity Measurements of Metallic Powders Under 1-Micron Wavelength Laser Light, presented at the 25th Annual International Solid Freeform Fabrication Symposium, Austin, TX, United States, 2014.

CHAPTER SIX

Equipment Qualification

Contents

1 DEFINITIONS OF TERMS

Calibration: the set of operations, which establish, under specified conditions, the relationship between values indicated by a measuring instrument or process and the corresponding known values of the measurand [1].

Critical (Direct Impact) utility: utility that is in direct contact with the product or that could have a direct impact on the quality of the product [2].

Critical process parameter: process parameter that is controlled within a predetermined range to ensure that the product meets its CQA.

Critical quality attributes (CQA): characteristics inherent in the product, that describe the acceptability of the products for use.

Installation qualification: documented evidence that the equipment, system or utility meets all critical installation requirements [3,4]

Operational qualification: documented evidence that the equipment, system or utility operates as intended throughout all required ranges.

Performance qualification: documented evidence that the equipment, system or utility performs as intended and meets all preestablished acceptance criteria

Standards, Quality Control, and Measurement Sciences in 3D Printing and Additive Manufacturing
http://dx.doi.org/10.1016/B978-0-12-813489-4.00006-4

Qualification: action of proving that any premises, systems and items of equipment are work correctly and actually lead to the expected results [5].

Validation: action of proving and documenting that the process consistently produces a product meeting its predetermined specifications and quality attributes [6].

2 INTRODUCTION TO QUALIFICATION

Qualification and validation of processes are necessary to ensure that the operation of processes falls within a predetermined optimal processing window to achieve sustainable commercial manufacturing. In many industries, qualification measures are required, especially in the medical and pharmaceutical manufacturing field. Food and Drug Administration (FDA) regulations, such as Current Good Manufacturing Practice (cGMP) for pharmaceuticals, GxP [such as Good Laboratory Practice (GLP), Good Clinical Practice (GCP)], Good Automated Manufacturing Practice (GAMP) guide published by International Society for Pharmaceutical Engineering (ISPE) [7], and industry standard ISO 9000, also require good documentation in establishing a controlled manufacturing process [8].

According to FDA's Guideline on General Principles of Process Validation, the concept of validation focuses on establishing documented evidence, which ensures a specific process will consistently produce a product meeting its predetermined specifications and quality attributes [6]. This concept is an essential part of cGMP. Each production process must be validated and therefore include all the important equipment used during the process. It is notable that these regulations do not provide specific instruction regarding the requirements of these qualification documentations. Individual manufacturing companies can design their own validation and qualification documentation system to suit their quality systems. Validation or qualification activities are carried out with careful planning of tests and acceptance criteria that are defined in advance. These criteria should be listed in a preapproved document called a protocol. Validation usually incorporates the concept of qualification. This means, in general, qualification can be viewed as a subset of validation. Fig. 6.1 shows the conceptual relationship between qualification and validation.

Qualification and validation are important in establishing and maintaining a controlled quality management system during manufacturing [9]. A validated manufacturing process can be monitored continuously through statistical process control approach to ensure product quality. Qualification and validation

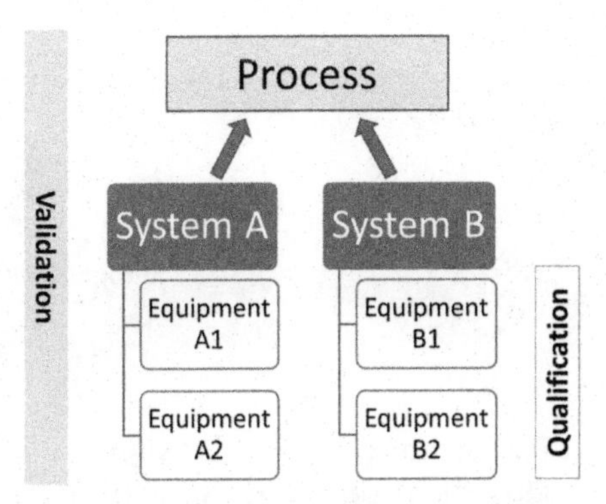

Figure 6.1 ***Qualification can be Viewed as a Subset of Activities During Validation.***

also support the implementation of continual improvement in equipment and manufacturing processes. As a result, the owner of the process will have increased process knowledge through performing qualification. Generally, process qualification includes the following aspects of manufacturing:

- facility,
- utilities,
- equipment,
- personnel,
- manufacturing process flow, and
- process monitoring.

In this chapter, the focus is on the qualification of equipment and utilities in an additive manufacturing (AM) process chain. It is noted that the validation of an AM process is not within the scope of this chapter. The validation of an AM process is still an on-going research effort and will require extensive knowledge beyond equipment qualification. It is the objective of this chapter, which concentrates on equipment, to provide a fundamental and standardized platform for the future development of a standardized AM process validation plan.

3 EQUIPMENT QUALIFICATION AND GENERAL TEST

Fig. 6.2 shows an example of a typical equipment qualification (EQ) plan [10]:

- Design Qualification (DQ).
- Installation Qualification (IQ).
- Operation Qualification (OQ).
- Performance Qualification (PQ).

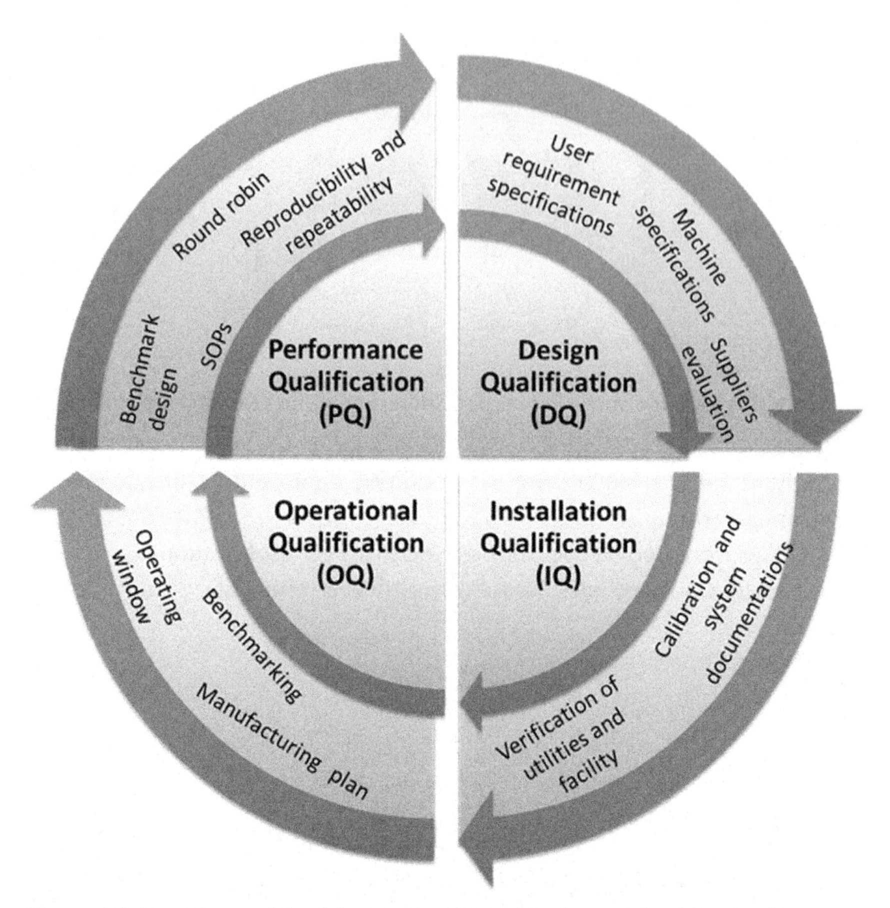

Figure 6.2 *Four Stages of Qualification in a Typical Equipment Qualification Plan.*

At each qualification stage, a protocol should be preapproved by the quality team. The protocol should clearly define the necessary tests to be conducted during each qualification stage along with the acceptance criteria. A report should be produced after each qualification stage. The reports document the results of tests conducted. The report should also document and addresses any out-of-specification results or nonconformances encountered during qualification activities.

3.1 Design Qualification

Design qualification (DQ) defines the functional and operational specifications of the equipment and details the conscious decisions in the selection of the supplier [1]. The equipment comes with the required functions and performance criteria to meet user requirements. A documented supplier's evaluation at this stage will also ensure that the supplier has sufficient supporting capability to provide training and support subsequent IQ. It is also important to ensure that the supplier is able to provide product support or software upgrade support throughout the life cycle of the machine.

The list belowshows the recommended steps that should be considered in a DQ.

1. Description of the intended use for the equipment and user requirement specifications (URS).
2. Description of functional, operational, and safety specifications.
3. Description of computer and software.
4. Evaluation of suppliers or vendors.

3.2 Installation Qualification

The installation qualification (IQ) establishes confidence that both hardware and software of the equipment fulfil the approved design intentions, and that the manufacturer's recommendations are suitably considered [6]. IQ establishes that the equipment is as specified and is properly installed. Integration with other equipment and utilities should also be verified at this stage. In IQ, the installation environment is also checked for stable and safe use of the equipment.

3.2.1 General IQ Test of AM Systems

1. Check the installation site to consider the manufacturer's recommendations.
2. Physical inspection of the machine.
3. Verification of system documentation including hardware and software. It could include technical data sheets, functional specification requirements of the machine and software, material of construction and certificates for critical subcomponents.
4. Verify safety data sheet of recommended printer materials.
5. Verify availability of equipment-related documentation, such as operating manuals, logbooks, and software installation disks.
6. Verify the list of spare parts, supporting utilities, and fulfilment of them.
7. Verify availability of calibration requirement that component is properly calibrated with documentation.
8. Define maintenance and calibration frequency with manufacturer's recommendation.

In the commercial landscape of AM systems, the vendors that supply the system will largely be able to assist in IQ. The vendor can provide IQ documentation and perform part of the functional testing as factory acceptance test at manufacturer's site. This will increase the efficiency and reliability of a well-planned IQ.

3.3 Operation Qualification

Operation qualification (OQ) is needed during scaling-up from product development to actual manufacturing equipment or when optimizing the process by testing different operating parameters. The OQ establishes confidence that both hardware and software of the equipment are capable of operating consistently within the established limits and tolerances. Critical process functions must meet the operating specifications and appropriately respond under faulty conditions. Hence it is necessary to test for the "worst-case scenario" and to provide documented demonstration that the equipment will perform as expected while operating at the extremes of the proposed range of operation. As the operating conditions for different materials or design features might vary in AM, it is therefore important to identify all the materials or intended applications that will be produced using the particular AM equipment. This will ensure that OQ can be performed in a realistic manner and to avoid overly extensive qualification testing.

3.3.1 General OQ Test of AM Systems

1. Verify that all OQ prerequisites are complete in that all IQ sections are completed or that any unexecuted sections would not impact the execution activities of the OQ.
2. Standard operating procedures (SOPs) are in place and approved and all required individual training is documented and available.
3. Verify that all subsystem or instruments that are critical to the operation of the system are calibrated.
4. Perform power failure and recovery tests and document the effects of these events on the control of the system.
5. Prepare a standard template as process control document with the important setup information of the document to be filled along with the finished parts.
6. Benchmark parts are designed specifically and suit the intended purpose of OQ. In-depth discussion of benchmark part design is presented in Chapter 8.
7. Determine the corresponding printing parameters and critical process parameters.

8. The critical quality attributes (CQA) of the part should be tested to cover the anticipated operating range of the machine, including "worst-case scenario".
9. Verify the standards test methods for measuring the CQA and characterization the attributes including different properties such as mechanical properties, tensile strength, elongation at break, surface quality, part density, and thermal properties.
10. Establish a manufacturing plan to standardize the printing operation during OQ. Examples of a manufacturing plan should include the following [11]:
    - part geometry, location in the build volume, and orientation,
    - raw material information and material handling,
    - building platform/chamber requirements,
    - machine setup, print parameters, such as print mode, laser exposure settings, and laser path strategy,
    - in process monitoring (if any),
    - part removal procedure, and
    - postprocessing procedure of the part.

A list of relevant standards or work items during OQ are given as follows:

- ASTM F2971-13 Standard Practice for Reporting Data for Test Specimens Prepared by Additive Manufacturing.
- ISO/ASTM52921-13 Standard Terminology for Additive Manufacturing-Coordinate Systems and Test Methodologies.
- ISO/ASTM52915-16 Standard Specification for Additive Manufacturing File Format (AMF) Version 1.2.
- ASTM F3122-14 Standard Guide for Evaluating Mechanical Properties of Metal Materials Made via Additive Manufacturing Processes.

Details of these standards can be found in Chapter 2.

3.4 Performance Qualification

Performance qualification (PQ) is the process of demonstrating that the equipment consistently performs as intended and meets all preestablished acceptance criteria. PQ is generally performed under conditions that represent the routine production conditions. PQ includes the integration of standard operating procedure, personnel, and systems, build plan, and materials to verify that the equipment consistently and repeatedly produces the required output [2].

3.4.1 General PQ Test of AM Systems

1. Approved standard operating procedures (SOPs), such as system operation, operation of test equipment, sampling plan, test method.

2. Raw material certification should be available.
3. The benchmark part design and build configuration should be representative of all anticipated production run scenario.
4. Calibration and maintenance requirement of the system should be verified.
5. Critical quality attributes of the parts are defined and characterized. Functional property of the benchmark part should be considered.
6. A preapproved and detail manufacturing plan with specification on the production sequence, machine and processing parameters, feedstock, postprocessing used and the measurement procedure used.
7. Precision and ruggedness of the manufacturing plan could be verified using a pilot run before PQ.
8. Critical process parameters are tested for reproducibility and repeatability (R&R).
9. Variations in the benchmark part should be focusing on between-operator, between-build, and between measurement effects.

PQ could be an extensive and expensive activity as a variety of processing parameters and variables is allowed in creating any AM parts [11]. It is, therefore a logical consensus to perform PQ as part of a round robin test to achieve a high level of confidence in specific AM equipment. A round robin study, or inter-laboratory study (ILS), is an established methodology to determine the reproducibility of a test method [11]. Analogies have been made to learn from round robin tests for analytical test methods as stated in ASTM E691 and ASTM E1169 to bring new values in the round robin testing of AM processes. Standard round robin study protocols were one of the priority action items identified in the Measurement Science Roadmap for Metals-Based Additive Manufacturing (see Chapter 2). A well-planned and well-executed round robin study will be able to quantify the repeatability and reproducibility (R&R) of a benchmark part and thus the system that was used to print that part. Details recommendations of Protocol for AM Round Robin Studies can be found in reference [11].

Following is a list of relevant standards or work items from ASTM for PQ:

- F2924-14 Standard Specification for Additive Manufacturing Titanium-6 Aluminum-4 Vanadium with Powder Bed Fusion.
- F3001-14 Standard Specification for Additive Manufacturing Titanium-6 Aluminum-4 Vanadium ELI (Extra Low Interstitial) with Powder Bed Fusion.
- F3049-14 Standard Guide for Characterizing Properties of Metal Powders Used for Additive Manufacturing Processes.

- F3055-14a Standard Specification for Additive Manufacturing Nickel Alloy (UNS N07718) with Powder Bed Fusion.
- F3056-14e1 Standard Specification for Additive Manufacturing Nickel Alloy (UNS N06625) with Powder Bed Fusion.
- F3091/F3091M-14 Standard Specification for Powder Bed Fusion of Plastic Materials.
- F3184-16 Standard Specification for Additive Manufacturing Stainless Steel Alloy (UNS S31603) with Powder Bed Fusion.
- F3187-16 Standard Guide for Directed Energy Deposition of Metals.
- WK55297 Additive Manufacturing—General Principles—Standard Test Artefacts for Additive Manufacturing.
- WK56649 New Guide for Standard Practice/Guide for Intentionally Seeding Flaws in Additively Manufactured (AM) Parts.
- WK49230 New Guide for Conducting Round Robin Studies for Additive Manufacturing.
- WK38342 New Guide for Design for Additive Manufacturing.
- WK54856 Principles of Design Rules in Additive Manufacturing.
- WK49229 New Guide for Orientation and Location Dependence Mechanical Properties for Metal Additive Manufacturing.
- WK49272 New Test Methods for Characterization of Powder Flow Properties for AM Applications.
- WK53878 Additive Manufacturing—Material Extrusion-Based Additive Manufacturing of Plastic Materials—Part 1: Feedstock materials Details of these standards can be found in Chapter 2.

4 CRITICAL UTILITIES

Critical utilities are needed during processing and for postprinting to support the whole AM process chain. One of the key steps within AM process flow is the postprocessing of the printed part and removal of support material [12]. This section will discuss the supporting utilities and identify the significance of these utilities. Functional information of these utilities will be necessary when deciding if these utilities need to be qualified.

The general utilities in a typical AM facility include gases, liquids and electrical systems.

Gases: the most common gases used in AM facilities are compressed air used for system operation or product cleaning and insert gas such as nitrogen and helium used for providing an inert environment in the build chamber.

Liquids: the most prevalent liquid utilities used are the recommended solvents that are used in cleaning operations and removal of support materials. The solvent directly impacts the cleaning efficiency of the printed part. It is therefore important to identify the risk of these liquids when establishing the qualification master plan.

Electrical: the electrical systems are necessary to power up the AM equipment. In critical operation, the power should be delivered continuously without stoppage. Some of the important electrical system attributes will include frequency, phase, and voltage requirements, as well as capacity. In some cases of the critical manufacturing operation, there might be a need to qualify an uninterruptible power supply (UPS) as part of the qualification plan.

4.1 Solid-Based Process-Specific Utilities

In the solid-based process, material extrusion system is used to illustrate the importance of storage and cleaning utilities.

4.1.1 Storage Utilities

A suitable storage chamber is necessary for storing the canisters of filament, which are not in use. Recommendations from the vendor include storing the filaments at suitable temperature and humidity levels. It has been reported that the filament might absorb environmental moisture and compromise the printing quality. This is because voids may be created during the extrusion process once filament with high moisture content is heated up [13]. Water absorption will lead to swelling of filament and also has an effect on the glass transition temperature, which will change the viscosity and flow property of the ABS [14]. Hence, a controlled environment is needed to maintain the predefined storage conditions. The temperature consistency and distribution profile within the dry box should be considered.

4.1.2 Cleaning Utilities

Ultrasonic cleaning bath is needed to remove soluble support material for the extrusion-based process. The support material that can be removed easily from the printed parts is first removed mechanically and manually. Strong alkaline solvent solution as the cleaning solvent is prepared by mixing the solvent powder into the water bath at specific ratio recommended by the vendor. Printed parts with partially removed support material are then soaked in the alkaline solvent to completely dissolve the remaining support material. Solution bath is then heated to recommended temperature to speed up the dissolving process. Different materials would require a different dissolving temperature. Hence, it is important to set the appropriate working temperature of the solution bath according to the type of material for consistent cleaning outcome. Ultrasonic and mechanical stirring are sometimes used to accelerate the dissolution process. The dissolving process normally takes a few hours to complete. Once the process is complete, the printed parts are removed from the solution bath and are then rinsed with water to wash away the alkaline solvent. Fig. 6.3 illustrates the cleaning processes of a part printed with FDM.

4.2 Liquid-Based Process-Specific Utilities

In the liquid-based process, material-jetting system is used to illustrate the importance of cleaning and postcuring utilities. Fig 6.4 illustrates the cleaning processes of a part printed with photopolymer jetting. In the process, there are multiple equipment and solvents involved.

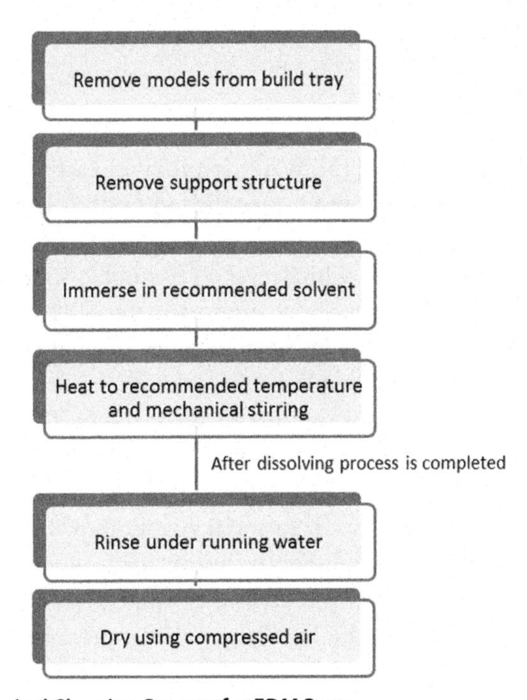

Figure 6.3 ***A Typical Cleaning Process for FDM Parts.***

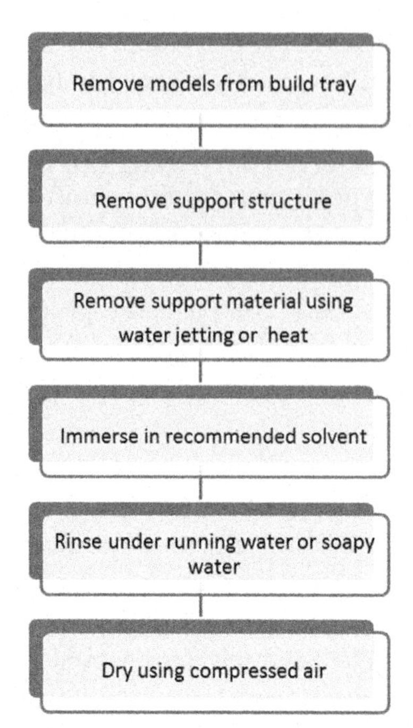

Figure 6.4 ***A Typical Cleaning Process for Material Jetting AM Parts.***

4.2.1 Cleaning Utilities

Waterjet station is needed to remove the water-soluble support material from PolyJet printed models [15]. Waterjet provides easy and fast cleaning of support material using high water pressure, without damaging the model materials.

In process, such as ProJet, which uses heat-soluble support material, a heating oven is used to melt the wax support material [16]. The wax support material should be melted by heating in an oven at predefined temperature and time to ensure complete removal. Fig. 6.5 shows printed parts are placed in a heating oven for support material removal.

In many cases, further cleaning is required. Vendors' recommended solvent such as sodium hydroxide or mineral oil would be used to remove fine traces of support material from the model material. An additional step of removing the mineral oil is then accomplished by using soapy water or IPA cleaning. An ultrasonic cleaner might be utilized to improve the cleaning operation. Ultrasonic wave agitates the liquid such that it produces high forces to remove support material or support structures adhering to the printed models. Compressed air is used to dry the parts after washing with water.

Figure 6.5 ***Bulk Support Material can be Removed by Heat in a Heating Oven.***

4.3 Powder-Based Process-Specific Utilities

4.3.1 In-Process Utilities

In laser melting processes, such as selective laser melting (SLM), inert gas flow, for example, argon or nitrogen is needed to maintain the atmosphere within the build chamber. The inert gas flow is needed to minimize contamination of the parts during high-temperature processing [17,18] and to remove the condensate produced during the melting process [19,20]. Condensate formed within the build chamber can result in reduction in the effective laser energy reaching the powder bed by absorption and laser scattering [20]. It has been shown that the flow rate, flow direction and type of gas will introduce changes to the resultant microstructure and porosity of the metal parts, which in turn, affects the mechanical properties of the fabricated parts [21,22]. Hence it is important to verify the effect of this utility during qualification of the laser-melting system. The temperature, pressure, flow rate, and capacity should be verified during the OQ.

4.3.2 Cleaning Utilities

In electron beam melting (EBM), compressed air is used to remove the powder cake after the process. Partially melted or sintered powder can then be removed and sieved for reuse [23]. A dehumidifier is needed to maintain the humidity of the powder bed. The humidity of the powder bed will subsequently affect the control of flowability of the powder [24]. In SLM, the parts built are usually buried in the powder bed, which requires removal by brushes after the process. Compressed air can then be used to remove any loose powder within the parts.

For the functional parts, they are attached to the build substrate directly or to the supports structures. With careful design of the support structures,

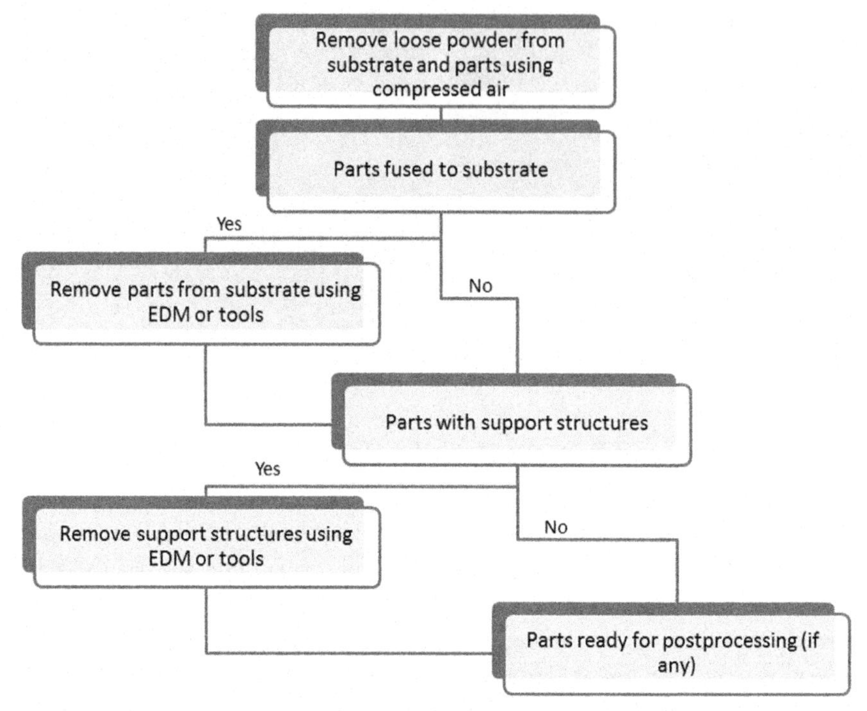

Figure 6.6 ***A Typical Cleaning Process for Powder Bed Fusion AM Parts.***

the parts can be removed by simple hand tools or postprocessing for removal will be required [25]. Removal of the parts and/or support structures can be done using electrical discharge machining (EDM), also commonly called wire cutting. Dielectric liquid or lubricating oil is normally needed during EDM operation. The manufacturer should consider the interaction of the fluid used during EDM as the fluid is contacting the part directly. Fig. 6.6 illustrates the cleaning processes of a part fabricated using powder bed fusion technologies.

5 EXISTING EQUIPMENT STANDARDS

Existing AM equipment standards are generally found within the domain of health and machinery safety. This section presents the common directives and standards, which the AM equipment are conformed to.

Directives:

- Machinery Directive 2006/42/EC.
- EMC Directive 2014/30/EU.
- RoHS Directive 2011/65/EU.

Machinery Directive 2006/42/EC is one of the essential health and safety requirements for machinery at EU level [26]. The Machinery Directive covers the safety aspects of machinery such as mechanical design, electrical design, controls, safety, and the potential for the machinery to create hazardous substances. The Electromagnetic Compatibility (EMC) Directive 2014/30/EU ensures that electrical and electronic equipment does not generate, or is not affected by, electromagnetic disturbance [27]. The Restriction of the Use of Certain Hazardous Substances (RoHS) in Electrical and Electronic Equipment Directive (2011/65/EU) aims to prevent hazardous substances from entering the production process and waste stream [28].

Some examples of specific standards for the safety of machinery design include:

- IEC 60825-1 Safety of laser products—part 1: equipment classification and requirements.
- IEC 62471-2 Photobiological safety of lamps and lamp systems.
- EN 13202 Ergonomics of the thermal environment—temperatures of touchable hot surfaces.
- EN 563 Safety of machinery—temperatures of touchable surfaces—ergonomics data to establish temperature limit values for hot surfaces.
- ISO 13732-1 Ergonomics of the thermal environment—methods for the assessment of human responses to contact with surfaces—part 1: hot surfaces.
- ISO 12100:2010 Safety of machinery—general principles for design—risk assessment and risk reduction.
- IEC 60204-1:2016 Safety of machinery—electrical equipment of machines—part 1: general requirements.

In summary, existing equipment standards applicable now are general standards, which are applicable to all machinery designs and not specific to AM equipment only. Thus, there is further potential to develop a specific roadmap of safety standards and regulations that AM equipment manufacturers should consider.

6 COMPUTER SYSTEMS AND SOFTWARE VALIDATION

Another important consideration in the qualification of AM equipment is the validation of computer system. All AM equipment is computer-linked systems for automatic and digital fabrication. The software applications in AM could include the use for coordinate transformation

[29], build-job consolidation [30], and process parameters prediction [31]. While it is not a prerequisite now to validate the operating software of the AM system, the principle of Computer System Validation (CSV) will contribute toward the total quality system in an AM manufacturing site. The quality of a printed part is directly related to the quality of the input file and hence the processing software of the printer. This section highlights the key considerations of CSV and electronic records, which are critical in AM.

CSV applies the definition of validation to a computer or computerized system. Following the principles of process validation, CSV requires similar qualification flow as follows:

- User to provide user-requirement specifications (URS).
- Supplier of the Computer System to provide design and functional specifications.
- IQ, OQ, and PQ of the computer system to verify that URS are met.

Regulatory and industrial bodies have also recommended using risk analysis techniques in evaluating the extensity of CSV. The complexity and the criticality of the software can be used to support and mitigate identified risk [32,33]. Other reference document in software validation includes ISPE GAMP5 [34]. The GAMP guide is a set of guidelines for companies in understanding and meeting cGMP regulations for automated systems. In ISPE GAMP 5, software is categorized as follows:

- Category 1: Infrastructure Software.
- Category 2: Nonconfigured products (including commercial off the shelf software (COTS) that is used as installed).
- Category 3: Configured products.
- Category 4: Custom applications.

A guidance document called *Guidance for Industry Part 11, Electronic Records; Electronic Signatures—Scope and Application* is also available from FDA to ensure compliant with electronic records and signature rules (21 CFR Part 11). Nonbinding recommendations in this guidance document are as follows [35]:

- Limiting system access to authorized individuals.
- Determination that persons who develop, maintain, or use electronic systems have the education, training, and experience to perform their assigned tasks.
- Establishment of and adherence to written policies that hold individuals accountable for actions initiated under their electronic signatures.
- Appropriate controls over systems documentation.

- Audit trails or other physical, logical, or procedural security measures in place to ensure the trustworthiness and reliability of the records.
- Requirements for generating copies of records and record retention.

In certain applications, the AM system can be controlled remotely on a networked environment. In this case, it could be necessary to qualify the network infrastructure as well. Testing should include access control to the network and stability of the network transactions under normal and high load. In addition to the computer system and software directly linked to the printers, it is also important to consider the preprocessing software of the part design file. With the design freedoms afforded by additive manufacturing (AM) processes, a variety of advances have emerged in topology optimization methods and data processing technologies [36]. This calls for renewed urgency to CSV to ensure the data file has been processed correctly and errors within a data file can be detected with traceability. The compatibility of the processed files with the printer's software is also a topic to be investigated.

In summary, this chapter has outlined methods to ensure that critical utilities, equipment and systems meet specifications, borrowing the concept of process validation in the regulated pharmaceutical industry. Unique considerations of AM processes are highlighted within the framework of equipment qualification (EQ) with reference to the commonly adopted stages of qualification (DQ, IQ, OQ, and PQ). The discussion presented here in this chapter serves as an initiation point to encourage further comprehensive and systematic investigation into the issue of AM equipment qualification.

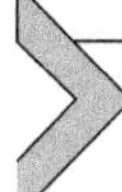

7 QUESTIONS

1. What are the differences between qualification and validation? Discuss the concept in terms of quality system management.
2. Name the important qualification stages of equipment qualification.
3. Why design qualification is important?
4. What is the purpose of Installation Qualification?
5. What are the key activities in Installation Qualification?
6. What are the key activities in Operational Qualification?
7. At what stage of qualification is a manufacturing plan needed?
8. What are included in a manufacturing plan in AM?
9. What are the challenges in Performance Qualification of an AM equipment?
10. What are the cleaning utilities in a material jetting process?

11. Is inert gas a critical utility in metal printing? Why?
12. What are the existing equipment standards that the manufacturers conform to?
13. Why computer system validation is applicable to AM equipment? What are the key considerations in performing software validation?

REFERENCES

[1] P. Bedson, M. Sargent, The development and application of guidance on equipment qualification of analytical instruments, Accred. Qual. Ass. 1 (1996) 265–274.

[2] Pharmaceutical engineering guides for new and renovated facilities, International Society for Pharmaceutical Engineering, 2001.

[3] 21 CFR: Parts 210 and 211—Current good manufacturing practice in manufacturing, processing, packing, or holding of drugs. United States Code of Federal Regulations, Food and Drug Administration, 1978.

[4] 21 CFR: Part 820—Good manufacturing practice regulations for medical devices, United States Code of Federal Regulations, Food and Drug Administration, 1996.

[5] WHO expert committee on specifications for pharmaceutical preparations—WHO technical report series, no. 961, World Health Organization, Switzerland, 2011.

[6] Guidance for industry process validation: general principles and practices, United States Code of Federal Regulations, Food and Drug Administration, 2011.

[7] Good Automated Manufacturing Practice Forum, Guide for Validation of Automated Systems in Pharmaceutical Manufacture Part 1 User Guide., ISPE, 1998.

[8] P.A. Cloud, Pharmaceutical Equipment Validation: the Ultimate Qualification Handbook, CRC Press, New York, (1998).

[9] W.Y. Yeong, C. K. Chua, Implementing additive manufacturing for medical devices: A quality perspective, presented at the 6th International Conference on Advanced Research and Rapid Prototyping (VRAP 2013), Leiria, Portugal, 2014.

[10] L. Huber, Validation and Qualification in Analytical Laboratories, Informa Healthcare, New York, (2007).

[11] S. Moylan, C.U. Brown, J. Slotwinski, Recommended protocol for round-robin studies in additive manufacturing, J. Test. Eval. 44 (2016) 1009–1018.

[12] C.K. Chua, K.F. Leong, 3D printing and additive manufacturing: principles and applications, 5th ed., World Scientific Publishing, Singapore, (2017).

[13] S.N.A.M. Halidi, J. Abdullah, Moisture and humidity effects on the ABS used in fused deposition modeling machine, Adv. Mater. Res. 576 (2012) 641–644.

[14] S.N.A.M. Halidi J. Abdullah, Moisture effects on the ABS used for fused deposition modeling rapid prototyping machine, presented at the IEEE Symposium on Humanities, Science and Engineering Research (SHUSER), Kuala Lumpur, Malaysia, 2012.

[15] Y.L. Yap, W.Y. Yeong, Shape recovery effect of 3D printed polymeric honeycomb, Virtual Phys. Prototyp. 10 (2015) 91–99.

[16] Y.L. Yap, C.C. Wang, H.K.J. Tan, V. Dikshit, W.Y. Yeong, Benchmarking of material jetting process: process capability study presented at the 2nd International Conference on Progress in Additive Manufacturing, Singapore, 2016.

[17] S.L. Sing, W.Y. Yeong, F.E. Wiria, Selective laser melting of titanium alloy with 50 wt% tantalum: Microstructure and mechanical properties, J. Alloys Comp. 660 (2016) 461–470.

[18] S.L. Sing, W.Y. Yeong, F.E. Wiria, B.Y. Tay, Characterization of titanium lattice structures fabricated by selective laser melting using an adapted compressive test method, Exp. Mech. 56 (2016) 735–748.

[19] B. Ferrar, L. Mullen, E. Jones, R. Stamp, C.J. Sutcliffe, Gas flow effects on selective laser melting (SLM) manufacturing performance, J. Mater. Process. Technol. 212 (2012) 355–364.

[20] A. Ladewig, G. Schlick, M. Fisser, V. Schulze, U. Glatzel, Influence of the shielding gas flow on the removal of process by-products in the selective laser melting process, Add. Manufactur. 10 (2016) 1–9.
[21] S. Dadbakhsh, L. Hao, N. Sewell, Effect of selective laser melting layout on the quality of stainless steel parts, Rapid Prototyp. J. 18 (2012) 21–249.
[22] A.B. Anwar, Q.-C. Pham, Selective laser melting of AlSi10Mg: effects of scan direction, part placement and inert gas flow velocity on tensile strength, J. Mater. Process. Technol. 240 (2017) 388–396.
[23] Y.H. Kok, X.P. Tan, N.H. Loh, S.B. Tor, C.K. Chua, Geometry dependence of microstructure and microhardness for selective electron beam-melted Ti–6Al–4V parts, Virtual Phys. Prototyp. 11 (2016) 183–191.
[24] A.T. Sutton, C.S. Kriewall, M.C. Leu, J.W. Newkirk, Powder characterisation techniques and effects of powder characteristics on part properties in powder-bed fusion processes, Virtual Phys. Prototyp. 12 (1) (2016) 3–29.
[25] M.X. Gan, C.H. Wong, Practical support structures for selective laser melting, J. Mater. Process. Technol. 238 (2016) 474–484.
[26] F. Flammer, The new machinery directive 2006/42/EC, Konstruktion (2008).
[27] M.W. Maynard, A new framework for the EU EMC directive the European Union new legislative framework and new EMC directive, in: 2015 IEEE Symposium on Electromagnetic Compatibility and Signal Integrity (EMC&SI 2015), Sillicon Valley, United States, 2015, pp. 7–11.
[28] O. Deubzer, N.F. Nissen, K.D. Lang, Overview of RoHS 2.0 and status of exemptions, presented at the Electronics Goes Green 2012+, ECG 2012—Joint International Conference and Exhibition, 2012.
[29] V.R. Shulunov, Algorithm for converting 3D objects into rolls using spiral coordinate system, Virtual Phys. Prototyp. 11 (2016) 91–97.
[30] J. Liu, Guidelines for AM part consolidation, Virtual Phys. Prototyp. 11 (2016) 133–141.
[31] C.Y. Yap, C.K. Chua, Z.L. Dong, An effective analytical model of selective laser melting, Virtual Phys. Prototyp. 11 (2016) 21–26.
[32] I.G. Forum, Risk assessment for use of automated systems supporting manufacturing processes: part 1—Functional risk, Pharma. Eng. 23 (2003) 16–26.
[33] G. Wingate, S. Brooks, Risk assessment for use of automated systems supporting manufacturing processes: part 2—risk to records, Pharma. Eng. 23 (2003) 30–40.
[34] H.Y. Charan, N. Vishal Gupta, GAMP 5: A quality risk management approach to computer system validation, Int. J. Pharma. Sci. Rev. Res. 36 (2016) 195–198.
[35] General principles of software validation; final guidance for industry and FDA staff, US Department of Health and Human Services, Food and Drug Administration, 2002.
[36] D.W. Rosen, A review of synthesis methods for additive manufacturing, Virtual Phys. Prototyp. 11 (2016) 305–317.

CHAPTER SEVEN

Process Control and Modeling

Contents

1 MOTIVATION FOR PROCESS CONTROL AND MODELING

Recently, additive manufacturing (AM) or 3D printing technology has caught the attention of the world [1–3]. This technology is deemed to be a paradigm shift in manufacturing, producing complex designs layer by layer from computer-aided design (CAD) file as compared to conventional subtractive manufacturing.

According to IDTechEx, metal printing is the fastest-growing segment in AM and it is believed that the upward trajectory will continue through 2020 [4]. In fact, more than 800 3D metal printers were sold in 2015—a growth of 46.9% over 2014 [5]. The main factor driving this growth in the market is the increasing use of metal printing processes in areas, such as aerospace, defense, medical, and dental.

Big names, such as GE Aviation, Siemens, NASA, Airbus, BMW, and Stryler are investing heavily in this technology to produce a myriad of

Standards, Quality Control, and Measurement Sciences in 3D Printing and Additive Manufacturing
http://dx.doi.org/10.1016/B978-0-12-813489-4.00007-6

components, such as fuel nozzles, blades for a gas turbine, rocket injectors, impeller wheels, and orthopedic implants [6]. The critical nature of these components requires stringent control of the manufacturing process. In order to be certified fit for use, there needs to be sufficient confidence that the components are free from any defects that may be induced during the AM process. The ideal method to overcome this concern is to be able to continuously monitor the AM process, identify causes of defects, make necessary in-process corrections, or amendments and provide data deterministic of the quality of the fabricated parts.

Effective in situ closed loop monitoring and feedback systems are deemed essential for increased adoption AM processes for industrial applications [7]. Through effective process monitoring, optimized parameters can be derived through analysis and interpretation of measurements obtained. This can greatly improve process reproducibility, as well as better assure quality and reliability of AM produced components.

Physics-based modeling and simulation are critical to predict the overall outcome of AM processes. It has great potential to improve part quality and therefore support the growth of AM. Not only does it explain the intrinsic physical interactions and the changes in the material phases and properties, it also offers greater insights into the process-parameter relationship.

Assuring the quality of additively manufactured parts to a similar degree as their conventionally manufactured equivalents is essential to enable large-scale adoption of AM. This chapter aims to demonstrate the increasing need for AM process control and modeling, which is a key requisite for the expected degree of part quality and assurance.

1.1 Process Parameters

Development of control and monitoring systems begins with the identification of all key process variables. Optimum process variables greatly improve the quality of produced parts. To develop an effective process control mechanism, extensive research is required to understand the physics governing the underlying process between each parameter. In AM, all core processes such as directed energy deposition (DED) [8], powder bed fusion (PBF) [9], vat photopolymerization [10], material jetting [11], binder jetting, material extrusion [12], and sheet lamination are inherently automated. Operators are only required to set up the machines with the appropriate materials, parameters, and print data files prior to the process.

Taking the example of a selective laser melting (SLM) process [13–16], there are four main categories that affect the ultimate quality of the finished parts and they are [7]:

- laser and scanning parameters,
- powder materials properties,
- powder bed properties and recoat parameters, and
- build environment parameters.

These four categories consist of many process variables that are largely correlated and interdependent. For example, the laser power is adjusted based on the different types of materials and scanning strategies used. In recent studies, these process variables are subdivided into two sections: predefined and controllable parameters.

Predefined parameters are parameters that are set and remain constant during the printing process. A few examples of predefined parameters are particle size distribution, material absorptivity, melting temperature, density, absorptivity, emissivity, etc. Controllable parameters commonly refer to scanning strategy, scanning velocity, layer thickness, pressure, gas flow velocity, ambient temperature, etc., which the user is allowed to tweak during the process in order to obtain the desired results.

The importance of understanding the basic process parameters can be demonstrated with the example of a SLM process, in which the laser absorptivity is dependent on the particle size and distribution of the metal powder. The smaller the size, the higher the absorptivity. Therefore, the controllable parameters, such as powder bed temperature, laser power, scanning speed and hatch spacing have to be optimized to achieve a balanced trade-off between melt-pool size, dimensional accuracy, surface finish, build rate, and desired mechanical properties. The build environment has to be maintained at a uniform temperature, pressure, and gas flow state to achieve repeatable results. Furthermore, modeling and simulation can be performed prior to such empirical studies which can greatly reduce the number of trial and error iterations required to determine the relationship between different process variables [17].

Using laser-based process as another example, higher laser power generally results in quicker melting of metal powder which produces much denser components. However, it may also lead to a large thermal variation across the powder bed which might, in turn, result in high residual stresses in the printed component [17–19]. While lower laser power helps to mitigate these residual stresses and assists in producing components of better geometrical accuracy, it may also result in parts that are of lower density and prone to delamination [17]. The choice of laser power to be used is deeply intertwined with desired spot size and scanning speed, scanning strategy, and platform temperature in order to produce quality parts [17–19]. For instance, melt-pool size, which is a key factor affecting final part quality, is dependent on a combination of laser power, spot size, scanning speed, scanning strategy, and platform temperature.

1.2 Process Parameter Resultants

Process parameter resultants refer to the qualitative output characteristics of a process, representative of the effect of certain input process parameters. The resultants can be classified as either observed or derived. Examples of observed resultants are melt-pool shape and size, spot temperature, and temperature distribution that can be measured on the spot through in situ process monitoring systems. Derived resultants include residual stress and melt-pool depth that can be derived through other means, such as modeling. Resultants provide a detailed understanding of the correlation between the process variables and the observable process characteristics, and the final product qualities, such as geometry, mechanical and physical characteristics. The measured values of these resultants are termed as the output variables of the process.

1.3 Control Objectives

Process control is a vital part of manufacturing. The three primary objectives of controlling an AM process are to:

- Optimize the process parameters: it is advantageous to assume some parameters, such as chamber pressure, gas flow, powder size, etc., to be constant. This will reduce the number of variables to control to achieve an optimized process.
- Increase the overall efficiency and repeatability: optimized parameters and machine settings are to be maintained and consistent, ensuring repeatability on the same print.
- Ensure the overall safety during the printing process: in general, for systems, such as PBF and DED, monitoring certain process parameters helps to foresee possible risks or hazards, such as fire or dust explosions. The equipment design could incorporate appropriate safety algorithms to reduce these risks, such as automatic powering off the laser beam if there is excess oxygen detected in the build chamber to reduce the risk of fire and turning off pressure pumps in case of abnormal back pressure due to clogged piping, and so forth.

National Institute of Standards and Technology (NIST), along with many other organizations and institutions, has identified the essential need of in situ process control for AM [20–22]. The value proposition of in situ process control lies in its ability to provide real-time visibility and control of the built environment. Through this, a continuous feedback system that is constantly engaged in analyzing the build and proactively correcting possible errors can be achieved. Ultimately, the goal is to enable systems with sufficient capability to qualify parts directly from the AM machine instead of postprocess inspection, as well as to enhance reliability and repeatability of the process.

Although much emphasis has been placed on the essentiality of a closed loop control system for AM, it is still in development stages. Practically, the challenges ahead are to achieve improvements in a feed-forward control system which enables better understanding of material properties, process parameters, performance, and their relationships to improve predictability.

In PBF or DED processes, the desire is to enable real-time data acquisition to monitor and control the melt-pool size and shape through corresponding process parameters. Fig. 7.1 illustrates a proposed framework that identifies the input parameters, and methods and sensors contributing to the output variables to achieve the control objectives or desired output.

According to the figure, control objectives are defined as the desired output, such as achieving temperature stability and consistency in melt-pool shape and size. For example, in order to maintain temperature stability in a melt-pool, one has to measure its temperature distribution (output variable) by using methods and sensors, such as photodiodes, IR cameras, charge-coupled device (CCD) camera, or complementary metal-oxide semiconductor camera (CMOS), and so forth. Likewise, input parameters, such as laser power, scanning speed, hatching speed, layer thickness, and material absorptivity will also impact the melt-pool temperature distribution which in turn contributes to its overall stability.

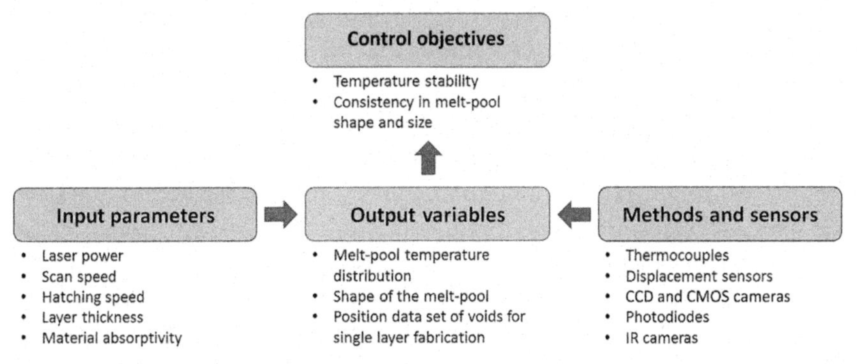

Figure 7.1 ***Control Objectives Framework.***

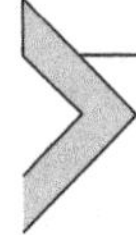

2 REVIEW OF STATE-OF-THE-ART RESEARCH ON IN-PROCESS CONTROL FOR ADDITIVE MANUFACTURING

This section consists of three subsections. The first explores some commonly used monitoring sensors that can be incorporated into AM systems. These sensors were used mainly to obtain process information and measurements, such as melt-pool size, melt-pool shape, spot size, temperature distribution, which were then analyzed to correspondingly vary the controllable parameters as part of their experiments. The second subsection delves into recent techniques and experimental setups which various researchers have applied on AM systems to obtain useful data. This portion focuses on four types of AM processes, namely SLM, electron beam melting (EBM), laser engineered net shaping (LENS), and wire-fed DED. The last subsection provides a summary of in-process control techniques.

2.1 Monitoring Sensors

In most AM processes, process parameters reflect a direct influence on part quality. In the case of SLM, process parameters, such as hatching space, scanning speed, laser power, material absorptivity, layer thickness, and build direction are directly associated with the properties of the printed parts, such as density, geometrical accuracy, surface roughness, and mechanical characteristics. It is known to be a typical scenario for most metal printing processes. During the process, a recoater mechanism applies a layer of powder evenly across the powder bed. Once the powder is deposited, a high-powered laser traces the cross-sectional geometry in accordance with the sliced stereolithography (STL) file, layer by layer, selectively melting, and fusing the material powder together. During this process, a significant amount of heat is generated. In metal printing processes, melt-pool temperature, and temperature distribution are attributed as the most critical factors during the printing process. Thus, achieving uniform controlled temperature distribution can bring about better microstructure, mechanical properties, geometrical accuracy, and surface finish [22]. Optimally controlling process parameters will result in better overall part quality if the related variables are identified, monitored, and properly modeled. Each process parameter that correlates to a process variable is required to be carefully monitored. In this section, different types of measuring instruments and sensors used to conduct experimental studies on monitoring and control methods are discussed.

2.1.1 Visual Imaging

A CMOS camera can be used to monitor the geometry of the melt-pool area, length, and width. The working principle of CMOS camera is relatively similar to a photodiode: it comes with a large array of photodiodes known as pixels that convert detected light to electric signals. The size of the melt-pool is measured by the pixel count captured by the camera. However, a CMOS camera is not the only type of sensor that is able to perform close-up monitoring.

A CCD camera is another type of monitoring camera that may be used to monitor geometry of the melt-pool area. Similar to a CMOS, a CCD camera has the ability to process detected light and convert it into electric signals. The primary difference between the two cameras lies in the number of processing microchips within it. A CCD camera uses a single processing unit to convert all the pixel signals, whereas each pixel in a CMOS has an individual charge to voltage convertor which offers a higher processing speed at the expense of greater complexity.

A CCD accumulates the photogenerated charges from each pixel and processes them using output circuitry. A CMOS converts light to voltage directly in each pixel [23]. In some cases, high pass filters are integrated into CCD and CMOS cameras for near infrared (NIR) imaging with the intention of capturing thermal intensity.

In a monitoring and control system, one may argue on the trade-off between the image resolution and processing speed. A CCD camera provides higher resolution, whereas a CMOS camera has a faster processing speed. Therefore, the choice of camera has to justify the intended purpose [23–25].

2.1.2 Thermal Sensing

A photodiode is commonly found in metal-based AM monitoring processes. This semiconductor device, which captures light intensity and converts it to electrical current, normally covers a small surface area. As the surface area increases, the response time of the device tends to be slower. The data captured by a photodiode in experiments generally refers to the melt-pool intensity in metal-based AM processes. The intensity detected by the sensor is proportional to the amount of light received.

Similar to a photodiode, a pyrometer is another device that may be used to measure thermal radiation emitted by the melt-pool. A pyrometer determines temperature based on measuring the intensity of IR radiation of a specific wavelength, as compared to a photodiode that measures the intensity of visible light emitted.

Apart from photodiodes and pyrometers, infrared cameras are also temperature monitoring equipment. Generally, they are also known as thermal imaging cameras. Similar to a pyrometer, an infrared camera is a noncontact device that detects thermal radiation. It converts thermal radiation into electric signals. The signals are then processed to produce thermal images, as well as temperature distribution plots for monitoring purposes.

The primary difference between the thermal devices depends on the region of interest. A photodiode or a pyrometer focuses on a single spot temperature measurement whereas an infrared camera is able to scan the temperature across a larger area. However, during large area scans with an infrared camera, critical thermal spikes are often missed out. This justifies the importance of single spot thermal reading with photodiodes or pyrometers [26,27].

2.1.3 Displacement Sensing

Apart from the cameras that capture images of the melt-pool, a displacement sensor can be used to measure the track height. This is especially beneficial in a DED process, which is also known by other trade names, such as LENS, direct metal deposition, laser cladding (LC) and laser metal deposition (LMD). In a DED process, a highly energized laser is used to melt the metal powders which are deposited onto a workpiece. The amount of metal powder deposited for DED has a direct impact on the geometry of the printed component. Hence, the track height is a useful measurement that can be obtained by a displacement sensor during such a process. The measured track height, along with the temperature profile across the melted track presents a strong correlation to the overall built quality.

The displacement sensor can be used to measure the displacement between the weld track and the reference position. There are two types of displacement sensors: noncontact and contact. The former uses optical, eddy current, ultrasonic, or laser devices to measure the displacement, while the latter uses a probe to directly contact the work piece. Noncontact optical displacement sensors and laser displacement sensors are the most commonly used types for monitoring the displaced track height. A research paper by Song et al. describes such a setup, where the optical displacement sensors were used to determine the track height, layer by layer, to ensure quality printing [28].

2.2 Measurand of In-Process Control Research in Additive Manufacturing

The majority of the studies and research related to in-process control are on metal-based processes where the focus lies mainly on melt-pool temperature and size, which are believed to be the most influential resultants affecting the overall quality of a printed product. This focus also helps to characterize the mechanisms behind the ability of metal-based AM processes to produce fully dense parts with mechanical properties that are equivalent to or better than bulk material. Moreover, this is critical for printing complex structures as the size and resolution of the melt-pool greatly affects the resolution of the print. In-process control is especially needed for aerospace components and medical devices with stringent quality control requirements. This section highlights on process control and monitoring techniques for metal-based AM and how sensors have been adopted to obtain the necessary data.

2.2.1 SLM

Kruth and Mercelis designed a patented feedback control system using a proportional-integral (PI) controller to control the laser power and stabilize the temperature distribution in the melt-pool [29]. In this setup, a high-speed CMOS camera and a photodiode were mounted coaxially to the laser beam. The photodiode was used to capture the light intensity of the melt-pool. This feedback control system was used during printing of test artifacts with geometrical features, such as overhangs. In their subsequent experiments based on this setup, they investigated on the influencing factors that affect the accuracy of the geometrical dimension, as well as identifying the possible process failures during the SLM process through in situ monitoring [30–32]. Recently, they introduced an image data processing algorithm to interpret the information obtain through the process [32]. Defects, such as deformation due to overheating at overhangs and thermal stresses are identified based on the images.

In a recent study by Clijsters et al., the melt-pool intensity was captured using a photodiode and a CMOS camera. The data was processed at a high sampling rate of 10 to 20 kHz for analysis. In order to generate a 3D model, the data was then mapped onto a grid, layer by layer, where each pixel from the grid represented a measured value of its position [33]. The results were then compared with X-ray computed tomography (XCT) images. The images from the generated 3D model drew high resemblance with the XCT images.

Yadroitsev et al. built a temperature monitoring system using a CCD camera coaxially aligned to the laser beam to monitor the melt-pool temperature distribution [34]. The focus of their research was to monitor the evolution of the microstructure during various stages of heat treatment. Chivel et al. developed a temperature monitoring system for SLM/SLS process using a CCD camera to monitor the process and a two-wavelength pyrometer for measuring the maximum surface temperature of the laser spot

[35]. Bayle et al. demonstrated their process monitoring technique using a high-speed IR camera and a pyrometer [36]. The work aims to obtain data relating to surface temperature, as well as the consolidation of the powder. However, the IR camera and pyrometer are not coaxially mounted, thus the quality of images obtained during the process may vary due to the viewing angle of the IR camera, as well as the position of the pyrometer. Similarly, Krauss et al. built a monitoring system using an IR camera to measure the temperature distribution across the entire workpiece instead of solely monitoring the melt-pool alone [37]. The authors aimed to investigate the lack of heat dissipation and other irregularities through the temperature distribution of the workpiece. Lott et al. designed a coaxial assembly consisting of a CMOS camera and an illumination source to obtain high-resolution images during scanning [38]. In the experiment, they managed to model the entire imaging light path using a ray tracing software.

2.2.2 EBM

SLM and EBM processes share a close resemblance. The main difference lies in the source of energy for fusion of metal powder and the built environment. EBM process uses an electron beam that selectively melts a bed of powder in a vacuum environment [39–41]. Unlike in a SLM process, it is extremely challenging to align the IR cameras to monitor the temperature distribution coaxially with the electron beam in an EBM process. The lack of space in the housing for the electron gun of the EBM system, greatly restricts the possibility of modifying the chamber to insert IR cameras. Mireles et al. developed an automatic closed loop system using an IR camera to provide layer by layer monitoring and feedback control for EBM process [42–44]. The authors aimed to maintain constant temperature during the creation of a controlled microstructure specimen. During the investigation, automatic control of process parameters, maintaining temperature stability, and detection of porosity during the process were carried out. Image processing was performed during the melting process by converting grayscale images into binary images. Two separate intensity thresholds were predefined by the user and compared to gauge temperature stability. Additionally, porosity identification was also carried out during the melting process. Dinwiddie et al. used an IR camera to monitor the printing process of overhanging structures. The authors acquired the images through the front shutter of the machine to monitor and detect critical phenomenas, such as porosity, overmelting issues during preheating and electron beam intensity measurement [45]. Price et al. used a NIR camera to monitor and measure the temperature distribution during the process [46,47]. In their work, they used two different lenses of different resolutions to monitor the behavior of the process parameters, such as build height effect, transmission losses due to metallization of sacrificial glass, overhanging structures, etc. Through the study, the temperature profile of the process captured using a NIR camera was found consistent and repeatable. In another work by Schwerdtfeger et al., an IR camera was used to monitor the printing process layer by layer. The IR camera images were then compared with those obtained through conventional metallographic sectioning [48]. Images from both metallographic sectioning and the IR camera showed a high resemblance for distribution of flaws.

2.2.3 DED

The working principle of a DED process is different from that of PBF in one fundamental way: the high power-density energy beam is focused on a continuous stream of powder or wire that is deposited onto the substrate, as opposed to a predeposited layer of powder. Bi et al. developed a closed loop system using a CCD camera coaxially aligned with the laser beam to capture the melt-pool, and a photodiode coaxially aligned with the nozzle head to capture the intensity of light emitted from the melt-pool [49]. They succeeded in obtaining a stable temperature distribution by varying the laser power. Similarly, Devesse et al. developed a closed loop system based on temperature monitoring [50]. In their setup, a NIR camera was used to measure the melt-pool surface temperature profile. The data was processed in real-time and sent directly to a controller. Notably, using a PI controller, the authors successfully controlled the melt-pool surface temperature profile obtained from the NIR camera by tweaking the laser power.

Köhler et al. developed a closed loop process using a CMOS camera and a pyrometer for measuring the melt-pool peak temperature. The authors thus modulated the laser power to keep the temperature constant along the cladding path [51]. In the experiment, the temperature field obtained from the sensors were evaluated and processed real-time. Additionally, the temperature field obtained during the process reciprocates with the results from finite element (FE) method. A monitoring system was developed by Smurov et al. using both a pyrometer and an IR camera to investigate the melt-pool and the heat affected zone (HAZ). In addition, a CCD camera was added to capture the powder distribution during the process to understand the chemistry between the gas flow and the powders which in turn enables optimization of powder injection [52]. Similar work by which the authors had captured the interaction of the powder using high-speed cameras in order to characterize the particle speed and flux can be found in refs [53,54]. Pekkarinen et al. built their monitoring system using a CCD camera and an illumination source to obtain high-quality images of the melt-pool [55]. The work highlighted the parametric study of laser power variation. Although Furumoto et al. also developed a similar monitoring system as Chivel et al. [35] to monitor surface temperature, their research mainly focused on controlling the consolidation pattern of the metal powder through the measurement of surface temperature by using a high-speed video camera and a two-color pyrometer to perform the measurements [56]. The two-color pyrometer has a higher sensitivity to the temperature change as compared to a two-wavelength pyrometer, thus accurately measuring the melting temperature range between 1520 and 1810°C.

Tang and Landers developed a feedback controller using a displacement sensor to measure the track height and a pyrometer to measure the melt-pool temperature [57]. The authors succeeded in achieving stable temperature control, consistent track width and height for a single track multilayer deposition. Multipass, multilayer deposition and height control could be integrated to obtain constant temperature. Song et al. presented a hybrid technique of controlling both melt-pool height and melt-pool temperature using three CCD cameras and a pyrometer [28]. It was concluded that the height of a layer is of higher importance as compared to the surface temperature. In the experiment, an algorithm was developed to change the laser power when the layer thickness fell below a defined limit. In other words, the temperature was not taken into consideration as long as the layer thickness remained within tolerable limits.

2.2.4 Wire-Fed DED

Wire-fed DED is another methodology that uses a wire as the primary material form to produce intricate geometry layer by layer. As a DED process, it uses either an electron beam or a laser to melt the material to create free-form parts. In this section, wire-fed electron beam process will be covered first, followed by a wire-fed laser process. Zalameda et al. employed a NIR camera to measure the temperature of the melt-pool and monitor the solidification of the area to produce quality builds [58] with the aim to adopt such technology on space related applications.

Liu et al. demonstrated monitoring of a wire-fed laser DED using a CCD camera to monitor the melt-pool and an illumination source to obtain high-quality images [59]. In addition, a spectrometer was used to closely monitor the emissivity of the plasma plume generated across the melt-pool. Conclusions, such as the coarsening of grain with the increase of voltage and laser power were drawn from the study. In another study by Heralic´ et al., a combination of two cameras was used—one to measure the width of the melt-pool from the top, coaxially aligned to the laser and another to measure the layer height [60]. In the setup, the temperature was not taken into consideration. The authors corrected the width and height of the weld track by varying the laser power to produce a stable process. Heralic´ et al. extended their research further by developing a control algorithm to integrate their 3D scanned data [61]. In their research, they managed to adjust the height and width of the weld track by the means of an iterative learning approach, producing a component with a high degree of accuracy.

2.3 Summary of In-Process Control and Monitoring Setup

This section provides a summary of the recently reviewed articles mentioned in Section 2.2. Most of the measuring techniques used in process control and monitoring utilize noncontact sensors. Table 7.1 summarizes

Table 7.1 Summary of in-process control and monitoring setups

Category	Type of process	Input variable	Measurand	Sensor	References
PBF	SLM/ EBM	Laser power/ beam current	Melt-pool temperature	CCD/CMOS camera	[29–33,35,38]
				IR camera	[36,37,42–48]
			Localized temperature	Pyrometer/ photodiode	[29–33,35,62]
DED	LENS/ laser wire-fed/ electron beam wire-fed	Laser power/ beam current	Melt-pool temperature	CCD/CMOS camera	[28,49,51,55,58,59]
				IR camera	[50,52]
			Localized temperature	Pyrometer/ photodiode	[28,49,51,52,56,57]
			Height	Displacement sensor	[57]
				CCD/CMOS camera	[28,60,61]

some existing in-process control and monitoring setups for different processes.

In the research reviewed, the temperature distribution was highlighted as the most critical factor in producing a quality part. In situ monitoring methods discussed in the papers were mostly focused on obtaining data related to the layer surface and the melt-pool. These reviews also highlight the importance of in-process control in producing parts of high quality and reliability. The data obtained from the feedback systems is necessary to change the input variable to achieve an ideal build environment. Detailed data acquisition mechanisms and comprehensive algorithms are necessary for enabling closed loop feedback. In summary, in situ monitoring and closed loop process control are keys to producing high-quality parts. In order to achieve success, a priori knowledge of process control is required.

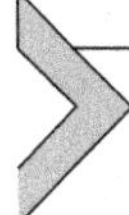

3 REVIEW OF STATE-OF-THE-ART ON PROCESS MODELING

Process models are mathematical abstractions of real processes. They enable the possibility to characterize the behavior of processes as long as their input parameters are known. The range of validity of these process models determines the situations in which they may be used.

Process models in AM are a cornerstone of any simulations or process control schemes. The accuracy of a simulation largely depends on the accuracy and comprehensiveness of the process model. Models are used for control of continuous processes, investigation of process dynamical properties, optimal process design, or for the calculation of optimal process working conditions.

Most state-of-the-art models for laser based AM processes in AM focus on these input parameters [19]:

- heat irradiating source (laser) characteristics,
- material domains in either voxel or particle,
- boundary conditions, and
- thermo-mechanical material properties.

The characteristics of a heat irradiating source, such as a laser beam include its associated power output, scan speed, pulse, and spot profile. The material domain is representative of the form of material, such as solid metal or powder. Boundary conditions are typically radiation and convection on the heated surface under adiabatic or isothermal conditions [19].

Process models may be broadly classified as either numeric models, which are typically modeled through multiphysics FE analysis, or analytical models, which are modeled with varying degrees of dimension, geometry, scale and varying phenomena, or subprocesses [19].

Typical FE methods of process modeling use voxel elements to represent solid metal domains. The energy source, usually a laser or electron beam, is modeled as a heat source with a Gaussian-shaped surface flux, consisting of variable beam diameter and power [63]. One method that utilizes particle domains for process modeling is the lattice Boltzmann method (LBM). Using particles instead of Navier–Stokes equation, this method can model physical phenomena that can include additional "uncontrollable" factors, such as powder density, stochastic effect of the powder bed, and so on. LBM allows the extraction of parameter-signature relationships, however, is computationally intensive and many simulations of the same model are required to achieve the desired outcome [64].

4 COMMERCIAL SOLUTIONS ON PROCESS MODELING IN ADDITIVE MANUFACTURING

As AM technology has already been in the market for a few decades, it is not surprising that there are multiple companies who have developed commercial softwares to model AM processes. Such softwares aims to achieve favorable results, as well as to reduce building time during actual printing. In this section, software packages by Simufact Engineering GmbH, ESI, and Autodesk Netfabb will be discussed. The software packages offered by these companies have the capability to predict defects, such as distortion and residual stresses. In addition, it also allows users to validate their printing strategy. In the last section, current commercial solutions for machines that are available in the market are discussed.

4.1 Simufact Additive

In 2016, Simufact Engineering GmbH, a software company based in Germany, launched Simufact Additive that provides a solution for the simulation of metal-based AM processes. This software is able to simulate metal-based AM processes, such as SLM and EBM, right from the start of the print job to subsequent post-processing steps, such as heat treatment and support structures removal.

The preliminary release of Simufact Additive incorporates the capability to simulate distortion and residual stresses in the printed metal parts. This comprises both a rapid mechanical approach for the prediction of distortion and residual stresses up to a fully thermo-mechanically coupled transient analysis. Temperature history and acquired properties, such as microstructure can be established and subsequently used for structural simulations.

The modeling is performed using CAD data through a graphical user interface environment oriented toward the real process workflow. This software is based on an intuitive approach in which the general process is first defined through the geometric design of the part and its support structures. The manufacturing parameters are then defined for analysis and results generation [65].

4.2 ESI-Additive Manufacturing

ESI has developed tools that focus on the heat source and powders interaction issues to identify defects, as well as residual stresses during the sintering process. The modeling solution provides distortion tools to predict the printed part's behavior during the build process and after removal from the machine. These tools are merged in a unified integrated computational material engineering (ICME) platform [66].

4.3 Netfabb Simulation

Simulations using Autodesk Netfabb allow for the prediction and alteration of deformations. This provides flexibility for designers and engineers to optimize designs and condense the iterative process necessary for consistent build results. The simulation also allows the user to validate the various build strategies [67].

Almost all parts presently produced by metal-based AM require postprocessing, such as machining in order to attain the desired final geometry and surface finish. The results obtained from the simulation further narrow down the process workflow and thus improve the overall efficiency of the manufacturing process. Additionally, simulation results can accurately determine the optimal process parameters necessary to produce the actual part, saving significant time with parameter optimization. Some of the features include automatic meshing through the use of hexahedral elements and mesh additivity. A multiscale modeling approach shortens the time required to complete the simulation process. Netfabb is also able to forecast potential failures during the printing in a powder bed processes which could avert damage to the equipment. Simulation using software tools, such as Netfabb helps to minimize potential damage incurred to the recoater blade, circumventing undesirable consequences, such as costly downtime and delays in production [67].

4.4 Current Commercial Process Control Solutions

Realization of in-process control is still considered to be in early development stage. Comparing DED and PBF, DED systems are more likely to achieve real-time control as the process speed is slower and the melt-pool is larger. In 2009, Optomec developed LENS MR7 that incorporates a closed loop melt-pool control system with thermal cameras that monitor temperature and cooling rate [68]. This system is used by the US Navy for various developmental works relating to repairs and creating new prototypes [69]. Sciaky Inc. developed a commercial interlayer real-time imaging and sensing system (IRISS) that is incorporated into their electron beam additive manufacturing (EBAM) systems. The IRISS provides closed loop control that enables EBAM machines to produce parts with high geometrical, mechanical and microstructural repeatability [70,71]. In 2016, Concept Laser GmbH developed the QMmeltpool 3D, a commercial monitoring system which aims to detect possible flaws, as well as provide real-time monitoring.

In recent years, there is a rising trend of third party companies developing such control and monitoring systems by incorporating their monitoring techniques into existing commercial machines. Sigma Labs Inc, Stratonics Inc and Plasmo Industrietechnik GmbH are a few companies that have developed such commercial capability, providing add-on process monitoring and control solutions for both DED and PBF processes [72–76]. However, the concern with such third-party solutions regarding the protection of intellectual properties remains a sensitive issue.

5 QUESTIONS

1. What is the control objective of a SLM process?
2. What are the commonly used sensors for in-process control?
3. In order to improve the quality of the images, what additional equipment can be employed in a visual monitoring system?
4. What are some of the current commercial solutions for process modeling and how do they benefit the AM research field?
5. List the most common input parameters for process models of laser-based AM processes.

REFERENCES

[1] C.K. Chua, K.F. Leong, 3D Printing and Additive Manufacturing: Principles and Applications, fifth ed., World Scientific Publishing Company, Singapore, (2017).

[2] C.K. Chua, M.V. Matham, Y.J. Kim, Lasers in 3D Printing and Manufacturing, World Scientific Publishing Company, Singapore, (2017).

[3] C.K. Chua, W.Y. Yeong, Bioprinting: Principles and Applications, World Scientific Publishing Company, Singapore, (2014).

[4] IDTechEx, 3D printing of metals 2015–2025. Available from: http://www.idtechex.com/research/reports/3d-printing-of-metals-2015-2025-000441.asp, 2015.

[5] T. Wohlers, 3D printing state of the industry. Available from: http://cgd.swissre.com/risk_dialogue_magazine/3D_printing/3D_Printing_State_of_the_Industry.html, 2016.

[6] M. K. Regan, Airplanes to medical devices, pioneering 3D printed titanium. Available from: http://blogs.ptc.com/2014/03/11/airplanes-to-medical-devices-pioneering-3d-printed-titanium/, 2014.

[7] J.-Y. Lee, W.S. Tan, J. An, C.K. Chua, C.Y. Tang, A.G. Fane, T.H. Chong, The potential to enhance membrane module design with 3D printing technology, J. Membr. Sci. 499 (2016) 480–490.

[8] J.S. Panchagnula, S. Simhambhatla, Inclined slicing and weld-deposition for additive manufacturing of metallic objects with large overhangs using higher order kinematics, Virtual Phys. Prototyp. 11 (2016) 99–108.

[9] W.S. Tan, C.K. Chua, T.H. Chong, A.G. Fane, A. Jia, 3D printing by selective laser sintering of polypropylene feed channel spacers for spiral wound membrane modules for the water industry, Virtual Phys. Prototyp. 11 (2016) 151–158.

[10] Y.Y.C. Choong, S. Maleksaeedi, H. Eng, P.-C. Su, J. Wei, Curing characteristics of shape memory polymers in 3D projection and laser stereolithography, Virtual Phys. Prototyp. (2016) 1–8.

[11] Z.X. Khoo, J.E.M. Teoh, Y. Liu, C.K. Chua, S. Yang, J. An, et al. 3D printing of smart materials: a review on recent progresses in 4D printing, Virtual Phys. Prototyp. 10 (2015) 103–122.

[12] M. Vaezi, S. Yang, Extrusion-based additive manufacturing of PEEK for biomedical applications, Virtual Phys. Prototyp. 10 (2015) 123–135.

[13] K.K. Wong, J.Y. Ho, K.C. Leong, T.N. Wong, Fabrication of heat sinks by Selective Laser Melting for convective heat transfer applications, Virtual Phys. Prototyp. 11 (2016) 159–165.

[14] W. Wu, S.B. Tor, C.K. Chua, K.F. Leong, A. Merchant, Investigation on processing of ASTM A131 Eh36 high tensile strength steel using selective laser melting, Virtual Phys. Prototyp. 10 (2015) 187–193.

[15] C.Y. Yap, C.K. Chua, Z.L. Dong, An effective analytical model of selective laser melting, Virtual Phys. Prototyp. 11 (2016) 21–26.

[16] C.Y. Yap, C.K. Chua, Z.L. Dong, Z.H. Liu, D.Q. Zhang, L.E. Loh, et al. Review of selective laser melting: materials and applications, Appl. Phys. Rev. 2 (2015) p.041101.

[17] I. Gibson, D. Rosen, B. Stucker, Additive Manufacturing Technologies, Springer Verlag, New York, (2015).

[18] W.J. Sames, F.A. List, S. Pannala, R.R. Dehoff, S.S. Babu, The metallurgy and processing science of metal additive manufacturing, Int. Mater. Rev. (2016) 1–46.

[19] M. Mani, B.M. Lane, M.A. Donmez, S.C. Feng, S.P. Moylan, A review on measurement science needs for real-time control of additive manufacturing metal powder bed fusion processes, Int. J. Prod. Res. (2016) 1–19.

[20] N.I.S.T. Energetics Inc, Measurement science roadmap for metal-based additive manufacturing, NIST (2013) 1–78.

[21] D.L. Bourell, M.C. Leu, D.W. Rosen, Roadmap for additive manufacturing - identifying the future of freeform processing, The University of Texas at Austin (2009) 2–92.

[22] K. Zeng, D. Pal, B. Stucker, A review of thermal analysis methods in laser sintering and selective laser melting, in Solid Freeform Fabrication Symposium Austin, TX, USA, 2012, pp. 796–814.

[23] D. Litwiller, CMOS vs. CCD: maturing technologies, maturing markets, Photonics spectra 39 (8) (2005) 54–61.

[24] D. Litwiller, CCD vs. CMOS: facts and fiction, Photonics spectra 35 (1) (2001) 154–158.

[25] Teledyne DALSA Inc, CCD vs. CMOS: which is better? It's complicated. Available from: http://www.teledynedalsa.com/imaging/knowledge-center/appnotes/ccd-vs-cmos/, 2013.
[26] FLIR Systems Inc, What is infrared? Available from: http://www.flir.com/about/display/?id=41528, 2016.
[27] Wikipedia, Thermogrpahic camera. Available from: https://en.wikipedia.org/wiki/Thermographic_camera, 2016.
[28] L. Song, V. Bagavath-Singh, B. Dutta, J. Mazumder, Control of melt pool temperature and deposition height during direct metal deposition process, Int. J. Adv. Manufact. Technol. 58 (2012) 247–256.
[29] J.-P. Kruth, P. Mercelis, Procedure and apparatus for in-situ monitoring and feedback control of selective laser powder processing, Google Patents, 2007.
[30] T. Craeghs, S. Clijsters, E. Yasa, F. Bechmann, S. Berumen, J.-P. Kruth, Determination of geometrical factors in layerwise laser melting using optical process monitoring, Opt. Lasers Eng. 49 (2011) 1440–1446.
[31] T. Craeghs, F. Bechmann, S. Berumen, J.-P. Kruth, Feedback control of layerwise laser melting using optical sensors, Phys. Procedia 5 (2010) 505–514.
[32] T. Craeghs, S. Clijsters, J.-P. Kruth, F. Bechmann, M.-C. Ebert, Detection of process failures in layerwise laser melting with optical process monitoring, Phys. Procedia 39 (2012) 753–759.
[33] S. Clijsters, T. Craeghs, S. Buls, K. Kempen, J.-P. Kruth, In situ quality control of the selective laser melting process using a high-speed, real-time melt pool monitoring system, Int. J. Adv. Manufact. Technol. 75 (2014) 1089–1101.
[34] I. Yadroitsev, P. Krakhmalev, I. Yadroitsava, Selective laser melting of Ti_6Al4V alloy for biomedical applications: temperature monitoring and microstructural evolution, J. Alloys Compd. 583 (2014) 404–409.
[35] Y. Chivel, I. Smurov, On-line temperature monitoring in selective laser sintering/melting, Phys. Procedia 5 (2010) 515–521.
[36] F. Bayle, M. Doubenskaia, Selective laser melting process monitoring with high speed infra-red camera and pyrometer, in: Fundamentals of Laser Assisted Micro-and Nanotechnologies, Proceedings of SPIE—The International Society for Optical Engineering (2008) p. 698505
[37] H. Krauss, C. Eschey, M. Zaeh, Thermography for monitoring the selective laser melting process, in: Solid Freeform Fabrication Symposium, Austin, TX, USA, 2012, pp. 999–1014.
[38] P. Lott, H. Schleifenbaum, W. Meiners, Design of an optical system for the in situ process monitoring of selective laser melting (SLM), Phys. Procedia 12 (2011) 683–690.
[39] L.E. Loh, Z.H. Liu, D.Q. Zhang, M. Mapar, S.L. Sing, C.K. Chua, et al. Selective Laser Melting of aluminium alloy using a uniform beam profile, Virtual Phys. Prototyp. 9 (2014) 11–16.
[40] Y. Kok, X. Tan, S.B. Tor, C.K. Chua, Fabrication and microstructural characterisation of additive manufactured Ti-6Al-4V parts by electron beam melting, Virtual and Phys. Prototyp. 10 (2015) 13–21.
[41] Y.H. Kok, X.P. Tan, N.H. Loh, S.B. Tor, C.K. Chua, Geometry dependence of microstructure and microhardness for selective electron beam-melted Ti–6Al–4V parts, Virtual Phys. Prototyp. 11 (2016) 183–191.
[42] J. Mireles, C. Terrazas, S.M. Gaytan, D.A. Roberson, R.B. Wicker, Closed-loop automatic feedback control in electron beam melting, Int. J. Adv. Manufactur. Technol. 78 (2015) 1193–1199.
[43] J. Mireles, C. Terrazas, F. Medina, R. Wicker, E. Paso, Automatic feedback control in electron beam melting using infrared thermography, Solid Freeform Fabrication Symposium, Austin, TX, USA, 2013, pp. 708-717.

[44] E. Rodriguez, F. Medina, D. Espalin, C. Terrazas, D. Muse, C. Henry, et al., Integration of a thermal imaging feedback control system in electron beam melting, in: Solid Freeform Fabrication Symposium, Austin, TX, USA, 2012, pp. 945-961.
[45] R.B. Dinwiddie, R.R. Dehoff, P.D. Lloyd, L.E. Lowe, J.B. Ulrich, Thermographic in-situ process monitoring of the electron-beam melting technology used in additive manufacturing, in: SPIE Defense, Security, and Sensing, 2013, pp. 87050K-9
[46] S. Price, J. Lydon, K. Cooper, K. Chou, Experimental temperature analysis of powder-based electron beam additive manufacturing, in: Solid Freeform Fabrication Symposium, Austin, TX, USA, 2013, pp. 162-173.
[47] S. Price, K. Cooper, K. Chou, Evaluations of temperature measurements by near-infrared thermography in powder-based electron-beam additive manufacturing, in: Solid Freeform Fabrication Symposium, Austin, TX, USA, 2012, pp. 761-773.
[48] J. Schwerdtfeger, R.F. Singer, C. Körner, In situ flaw detection by IR-imaging during electron beam melting, Rapid Prototyp. J. 18 (2012) 259–263.
[49] G. Bi, A. Gasser, K. Wissenbach, A. Drenker, R. Poprawe, Characterization of the process control for the direct laser metallic powder deposition, Surf. Coat. Technol. 201 (2006) 2676–2683.
[50] W. Devesse, D. De Baere, M. Hinderdael, P. Guillaume, Hardware-in-the-loop control of additive manufacturing processes using temperature feedback, J. Laser Appl. 28 (2016) 1–8.
[51] H. Köhler, V. Jayaraman, D. Brosch, F.X. Hutter, T. Seefeld, A novel thermal sensor applied for laser materials processing, Phys. Procedia 41 (2013) 502–508.
[52] I. Smurov, M. Doubenskaia, A. Zaitsev, Comprehensive analysis of laser cladding by means of optical diagnostics and numerical simulation, Surf. Coat. Technol. 220 (2013) 112–121.
[53] P. Balu, P. Leggett, R. Kovacevic, Parametric study on a coaxial multi-material powder flow in laser-based powder deposition process, J. Mater. Process. Technol. 212 (2012) 1598–1610.
[54] I. Smurov, M. Doubenskaia, A. Zaitsev, Complex analysis of laser cladding based on comprehensive optical diagnostics and numerical simulation, Phys. Procedia 39 (2012) 743–752.
[55] J. Pekkarinen, V. Kujanpää, A. Salminen, Laser cladding with scanning optics: effect of power adjustment, J. Laser Appl. 24 (2012) 032003.
[56] T. Furumoto, T. Ueda, M.R. Alkahari, A. Hosokawa, Investigation of laser consolidation process for metal powder by two-color pyrometer and high-speed video camera, CIRP Annals Manufact. Technol. 62 (2013) 223–226.
[57] L. Tang, R.G. Landers, Melt pool temperature control for laser metal deposition processes—part I: online temperature control, J. Manufact. Sci. Eng. 132 (2010) 011010.
[58] J.N. Zalameda, E.R. Burke, R.A. Hafley, K.M. Taminger, C.S. Domack, A. Brewer, et al., Thermal imaging for assessment of electron-beam freeform fabrication (EBF3) additive manufacturing deposits, in: SPIE Defense, Security, and Sensing, 2013, p. 87050M-8.
[59] S. Liu, W. Liu, M. Harooni, J. Ma, R. Kovacevic, Real-time monitoring of laser hot-wire cladding of inconel 625, Opt. Laser Technol. 62 (2014) 124–134.
[60] A. Heralic´, A.-K. Christiansson, M. Ottosson, B. Lennartson, Increased stability in laser metal wire deposition through feedback from optical measurements, Opt. Lasers Eng. 48 (2010) 478–485.
[61] A. Heralic´, A.-K. Christiansson, B. Lennartson, Height control of laser metal-wire deposition based on iterative learning control and 3D scanning, Opt. Lasers Eng. 50 (2012) 1230–1241.
[62] C.K. Chua K.F. Leong, 3D printing and additive manufacturing: principles and applications: World Scientific Publishing Company 2014.
[63] R.B. Patil, V. Yadava, Finite element analysis of temperature distribution in single metallic powder layer during metal laser sintering, Int. J. Mach. Tools Manufact. 47 (2007) 1069–1080.

[64] C. Körner, E. Attar, P. Heinl, Mesoscopic simulation of selective beam melting processes, J. Mater. Process. Technol. 211 (2011) 978–987.
[65] Metal-AM, Simufact to launch process simulation software solution for metal additive manufacturing. Available from: http://www.metal-am.com/simufact-launch-process-simulation-software-solution-metal-additive-manufacturing/, 2016.
[66] ESI-Additive manufacturing, Using simulation to model metallic additive manufacturing processes. Available from: https://www.esi-group.com/software-solutions/virtual-manufacturing/additive-manufacturing, 2016.
[67] Autodesk Netfabb, Autodesk launches Netfabb 2017 solution for additive manufacturing. Available from: https://www.netfabb.com/blog/autodesk-launches-netfabb-2017-solution-additive-manufacturing, 2016.
[68] OPTOMEC, LENS MR-7 systems. Available from: http://www.optomec.com/3d-printed-metals/lens-printers/metal-research-and-development-3d-printer/, 2016.
[69] Industrial Laser Solutions, LENS MR-7 system to be used for next generation laser additive manufacturing. Available from: http://www.industrial-lasers.com/articles/2009/10/lens-mr-7-system-to-be-used-for-next-generation-laser-additive-manufacturing.html, 2016.
[70] Sciaky Inc, Make metal parts faster & cheaper than ever with electron beam additive manufacturing (EBAM™) systems or services. Available from: http://additivemanufacturing.com/2015/08/24/endless-possibilities-with-sciakys-expanded-lineup-of-electron-beam-additive-manufacturing-ebam-systems/, 2017.
[71] AMazing, Endless possibilities with Sciaky's expanded lineup of electron beam additive manufacturing (EBAM) systems. Available from: http://additivemanufacturing.com/2015/08/24/endless-possibilities-with-sciakys-expanded-lineup-of-electron-beam-additive-manufacturing-ebam-systems/, 2017.
[72] C. Scott, Sigma labs releases latest version of PrintRite3D INSPECT software, based on early adopter feedback. Available from: https://3dprint.com/147115/sigma-labs-printrite3d-inspect/, 2016.
[73] Sigma Labs, Process control and quality assurance software for additive manufacturing. Available from: https://www.sigmalabsinc.com/products, 2016.
[74] M. Molitch-Hou, For Stratonics and metal 3D printing, the heat is on. Available from: http://www.engineering.com/AdvancedManufacturing/ArticleID/13133/For-Stratonics-and-Metal-3D-Printing-the-Heat-Is-on.aspx, 2016.
[75] Stratonics Inc, Sensors. Available from: http://stratonics.com/systems/sensors/, 2016.
[76] EOS GmbH, 2014. EOS and plasmo join forces in the field of online process monitoring for additive serial manufacturing. Available from: https://www.eos.info/eos_plasmo_online_process_monitoring_for_additive_manufacturing

CHAPTER EIGHT

Benchmarking for Additive Manufacturing

Contents

1 DESIGN OF TEST ARTIFACTS

Test artifacts have been used to either evaluate the individual processes or to compare between processes in order to determine the suitability of the processes for various applications. The test artifact should contain certain features that allow the quantification of several qualities of the process. A sample of a simple test artifact or benchmark is shown in Fig. 8.1.

Standards, Quality Control, and Measurement Sciences in 3D Printing and Additive Manufacturing
http://dx.doi.org/10.1016/B978-0-12-813489-4.00008-8

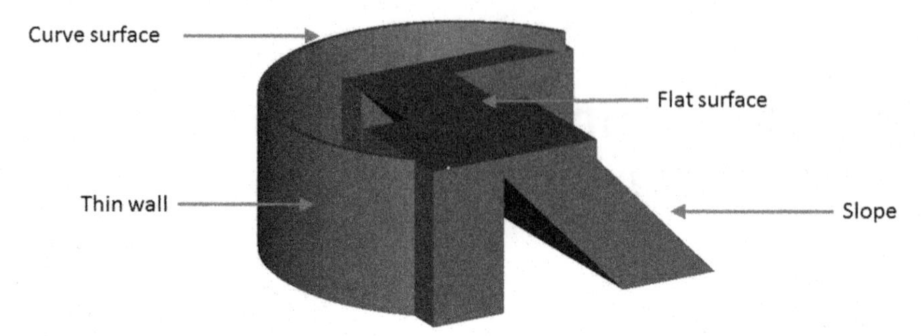

Figure 8.1 ***Sample of a Test Artifact.***

1.1 Considerations

In designing an artifact, a standard benchmark should not only provide ways to highlight errors and/or limitations of a machine or process, but also be able to correlate the errors and limitations to the specific aspects. The test artifact can also include replicates of the same feature to allow repeatability and test the capability of the process in producing the same features at different locations. However, this does not test the overall repeatability of the process.

Key features that should be able to be manufactured by AM are identified, such as [1]:

- straight features,
- parallel and perpendicular features,
- circular and arc features,
- fine features,
- freeform features,
- holes and bosses, and
- overhangs.

1.2 Benchmarking Individual Processes

Whenever there is a new process or material in the market, there is a need for benchmarking to ensure that there is process optimization or improvement [2–4]. Hopkinson and Sercombe used an artifact comprising multiple stair steps to study the accuracy of selective laser sintering (SLS) [5]. Campanelli et al. used a designed artifact to find the optimized parameters for dimensional accuracy of stereolithography (SL). In their artifact, a substantial number of small and medium dimensions are included to simulate small applications such as those in jewelry industry. Protrusions and cavities are

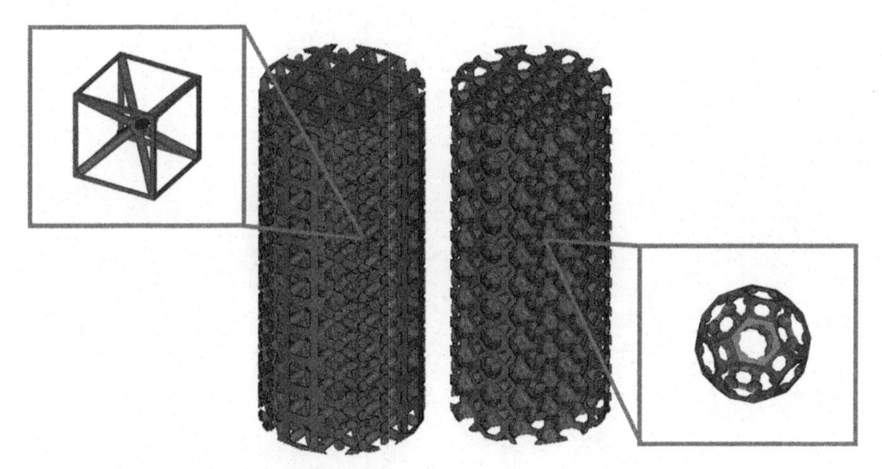

Figure 8.2 ***Lattice Structures as Benchmarks.***

used for the dimensional accuracy evaluation, whereas planes with different angles and overlapped cylinders are used for determination of position errors such as inclination and perpendicularity. Cylinders are also included to measure conicity [6]. Sing et al. used two different lattice structure designs as artifacts, as shown in Fig. 8.2, to study the limitation of selective laser melting (SLM) process in producing thin struts [7].

Yap et al. designed several benchmark artifacts to identify the optimized parameters for inkjet printing in terms of dimensional accuracy for various thicknesses and heights [8]. One example of the benchmark part for studying the limitation of inkjet printing in producing small features and thin walls is shown in Fig. 8.3.

In addition, benchmark artifact can also be designed to examine the geometrical limitation of assembly-free parts printed using inkjet printing by removing the unfused parts after postprocessing, as shown in Fig. 8.4.

1.3 Comparison Between Processes

With the increase in capability and acceptance of AM, there is an increase in the number of machine platforms by different manufacturers for the same process. This leads to the need for comparison of process capability between them.

Childs and Juster designed an artifact to test the geometrical capabilities, tolerances, limits and repeatability of different processes, such as selective laser sintering (SLS), laminated object manufacturing (LOM), and fused deposition modeling (FDM). The part includes a square base that allowed the measurement of flatness, straightness and right angles. Cylinders were also

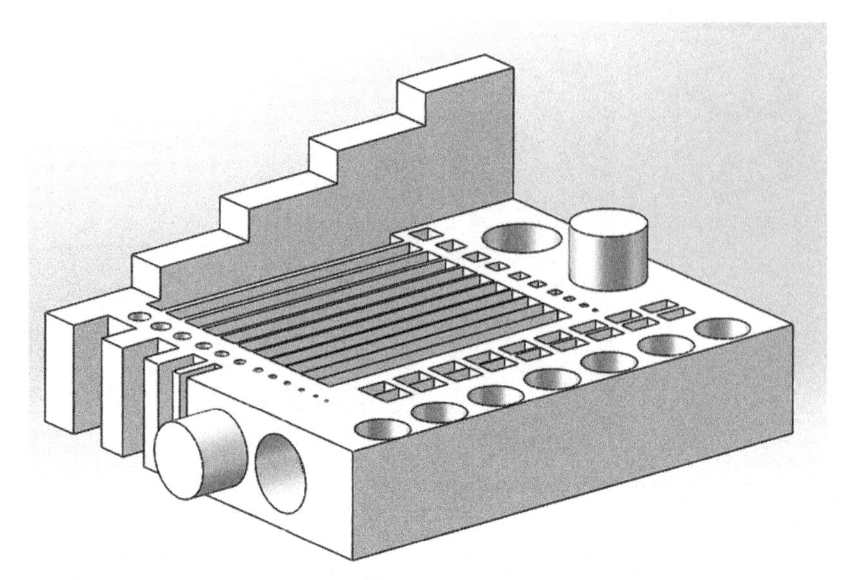

Figure 8.3 ***Benchmark Artifact for Characterizing Small Features and Thin Walls.***

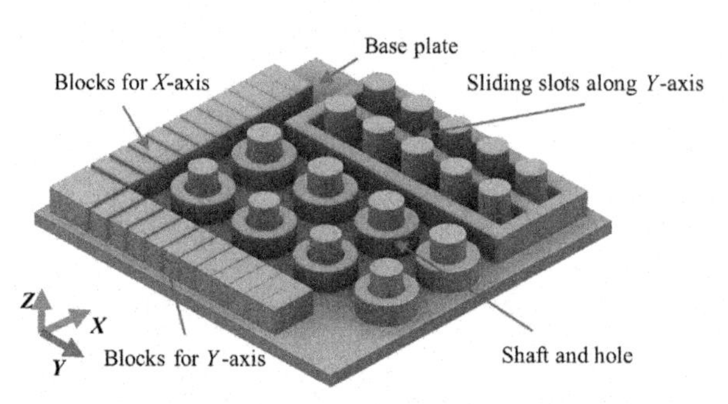

Figure 8.4 Benchmark Artifact to Examine the Minimum Clearance Required for Removing the Unfused Parts.

included to measure concentricity [9]. Xu et al. used a customized artifact to study the dimensional accuracy and surface finish of parts manufactured using different processes such as stereolithography, SLS, FDM, and LOM. Features, such as fine structures, overhangs, large flat surfaces, and small gaps are included [10]. The same processes were compared by Mahesh et al., however, using a different benchmark that includes new geometrical features such as freeform surfaces and pass-fail features [11,12]. The artifact was also used to compare between processes of similar principle such as direct metal laser sintering (DMLS), SLM and SLS. Abdel Ghany and Moustafa used a half die for glass bottle as a benchmark. It included complicated features such as fine holes, cooling channels, text with sharp edges and corners, fillets, chamfers, and thin walls [13]. Lee et al. have designed a benchmark specifically for microfluidic chip application to compare the printing resolution, accuracy, repeatability, circularity, surface roughness, and water contact angle of PolyJet and FDM printed benchmark artifact [14]. In their work, the intention is to characterize the process capability of ink-jet printing process in term of dimensional accuracy and surface quality as a function of printing orientation. NIST has also published a standard benchmark artifact to investigate the performance and capabilities of SLM and 3D printing processes [15]. The test artifact design was also proposed for formal standardization through the ASTM F42 committee on additive manufacturing.

2 METROLOGY MEASUREMENT METHODS

This section discusses about some of the measurement methods of the physical features of geometrical products. The list further shows some of the features commonly reported in AM parts:

- straightness, roundness, and roughness,
- porosity and density, and
- dimensions.

The description on the standards and measurement methods for each feature will be discussed in the following subsections.

2.1 Straightness, Roundness, and Roughness

The standards developed to guide the measurement of straightness, roundness, and roughness of a part are as follows:

ISO 12780 Geometrical product specifications (GPS)—straightness. This standard discusses about the concept of straightness, which belongs to one of the geometrical product specifications. This standard consists of two parts. The first part (ISO 12780-1:2011) covers the fundamental concept of straightness and vocabulary/terms for describing straightness, whereas the second part (ISO 12780-2:2011) talks about the specifications operators. ISO 12780-1 discusses about the profiles, reference line, and filter function. ISO 12780-2 discusses about the complete specification operator, such as the selection of appropriate transmission band according to the probing system.

ISO 12181 Geometrical product specifications (GPS)—roundness.

This standard discusses about the concept of roundness, which belongs to one of the geometrical product specifications. This standard consists of two parts. In the first part (ISO 12181-1:2011), the fundamental concepts of roundness and vocabulary/terms for describing roundness are covered. In the second part (ISO 12181-2:2011), the specifications operators are detailed. ISO 12181-1discusses about the profiles, reference circle, and filter function. ISO 12781-2 discusses about the complete specification operator, such as the selection of appropriate transmission band according to the probing system.

ASTM D7127 Standard test method for measurement of surface roughness of abrasive blast cleaned metal surfaces using a portable stylus instrument.

This standard describes the appropriate use of stylus surface profilometer to evaluate surface parameter and some considerations related to setup of the stylus instrument for data acquisition. It discusses about the evaluation length, sampling length, calibration of the instrument and the calculation of the surface parameters.

2.2 Porosity and Density

Density of a porous material may be defined as either apparent density or true density [16]. Apparent density Q_a is defined as the mass of material per unit external volume, whereas true density Q_t is defined as the mass per unit real volume [16]. The external volume includes the pores associated with the given mass of material, whereas real volume excludes any pores in the mass of the material [16]. The porosity ε can then be calculated using [16]:

$$\varepsilon = 1 - \frac{Q_a}{Q_t}$$

Generally, the true density of a material is measured by applying Archimedes Principle, which is by water displacement method using a pycnometer. However, it is important to saturate the material with wetting liquid, otherwise the displaced volume does not represent the real volume of the material. Alternatively, the real volume can be measured using X-ray diffraction method [16].

There are two simple ways to measure the apparent density of a porous material. The first method is by using a nonwetting liquid for the liquid displacement method [16]. One disadvantage of this method is the hydrostatic pressure tends to force the liquid into the pores, which will result in smaller external volume estimation. The second method is by applying an impermeable coating to the porous material; however, it is difficult to ensure no coating is sucked up into the pores [16]. ASTM B962 describes a test method for measuring porous material such as powder metallurgy products [17].

ASTM B962 Standard Test Methods for Density of Compacted or Sintered Powder Metallurgy (PM) Products Using Archimedes Principle.

This standard provides a method for determining the density of powder metallurgy products that usually have surface-connected porosity. This standard is suitable for green, compact, and sintered parts. This standard describes about the standard test procedures, apparatus and materials required for density measurement. As interference could occur when porous part is immersed in water bath, therefore, the porous part is first immersed in oil bath to prevent filling the gap. The density of the test specimen can then be calculated using the basic formula provided in the document.

2.3 Dimensions

Dimensions of the parts can be measured simply using Vernier calipers and rulers. For better precision, typically, measurement can be taken with magnified images from either, optical microscopy (OM), scanning electron microscopy (SEM), or X-ray microtomography (μCT). For direct measurement of the parts, a coordinate-measuring machine (CMM) can be used. OM uses visible light or laser and a system of lenses with different magnification to capture images of small samples. An example of an OM image of metal part is shown in Fig. 8.5.

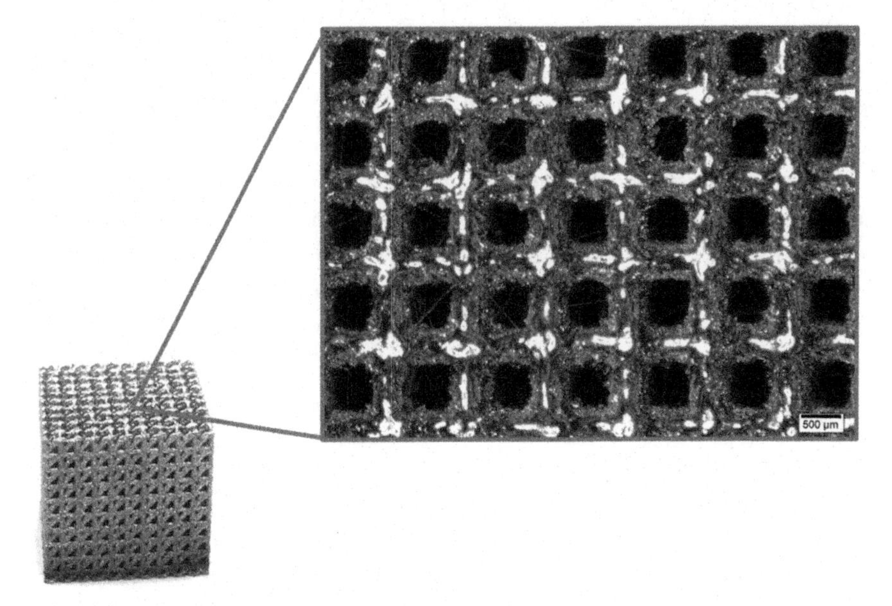

Figure 8.5 ***Optical Microscopy Image of Metal Part.***

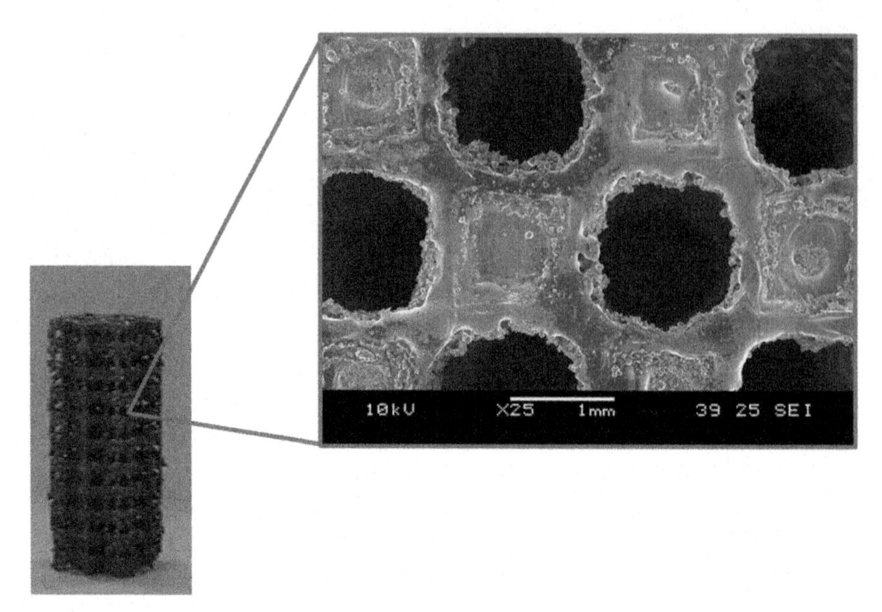

Figure 8.6 ***SEM Image of Metal Part.***

SEM produces images of a sample by scanning it with a focused beam of electrons. The images are obtained when the electrons interact with the atoms of the samples and produces signals that is detected together with the original beam position. An example of a SEM image of metal part is shown in Fig. 8.6.

CMM is a common measuring modality in precision engineering. It uses a probe, that can be mechanical, optical or laser, to obtain the location of a point of the part in the three axes, relative to the predetermined origin. Using these precise coordinates, points are generated, which can then be analyzed using regression algorithms to get reconstructed features or measurements. μCT uses X-rays to obtain cross-sections of a part, which then is used for reconstruction of a virtual three-dimensional (3D) model. The reconstructed model can then be used for measurements.

3 MECHANICAL MEASUREMENT METHODS

Material properties in AM can be generally found in the data sheets given by suppliers, as well as from individual testing. Mechanical properties of the printed parts allow engineers to choose the appropriate material for different applications. This section presents some mechanical properties that are commonly tested and reported for AM materials. It is the objective of this section to introduce the test without going into details of the theoretical mechanics of these test methods. It is also noted that the testing results shown are not the discussion of this section. This section serves to present the current commercial and research landscape in mechanical measurement methods of AM polymers and metals. Relevant standards test methods are listed and typical parameters reported for those tests are also presented. The readers are encouraged to refer to the listed standards and references for more detailed derivation of the test methods [18,19].

3.1 Polymers

Commonly tested mechanical properties of AM polymers are:
- tensile properties,
- compressive properties,
- flexural properties, and
- impact strength.

In addition, for liquid-based materials, the following properties are of critical importance to ensure the performance of the print part:
- water absorption and
- density as printed.

The details of the tests are explained in the following sections.

3.1.1 Tensile Properties Measurement Methods

Tensile properties show how the material will react to forces being applied in tension. Tensile tests are able to show the ability of a material to withstand tensile loads without failure. They can also measure the ability of a material to deform under tensile stresses. Tensile tests of polymers are used to determine the tensile strength, tensile modulus, tensile strain, and elongation at yield, or at break. These properties are important to determine if the material is suited for specific applications or if it will fail under specific stresses.

Test methods for measuring tensile properties of plastics are given by the following standards:

ASTM D638 Standard test method for tensile properties of plastics.

ISO 527-2 Plastics—determination of tensile properties—part 2: test conditions for molding and extrusion plastics.

Both standards provide varying dog-bone shaped specimen geometries and testing requirements for different type of materials to be tested. Fig. 8.7 shows a sample dog-bone tensile coupon being tested.

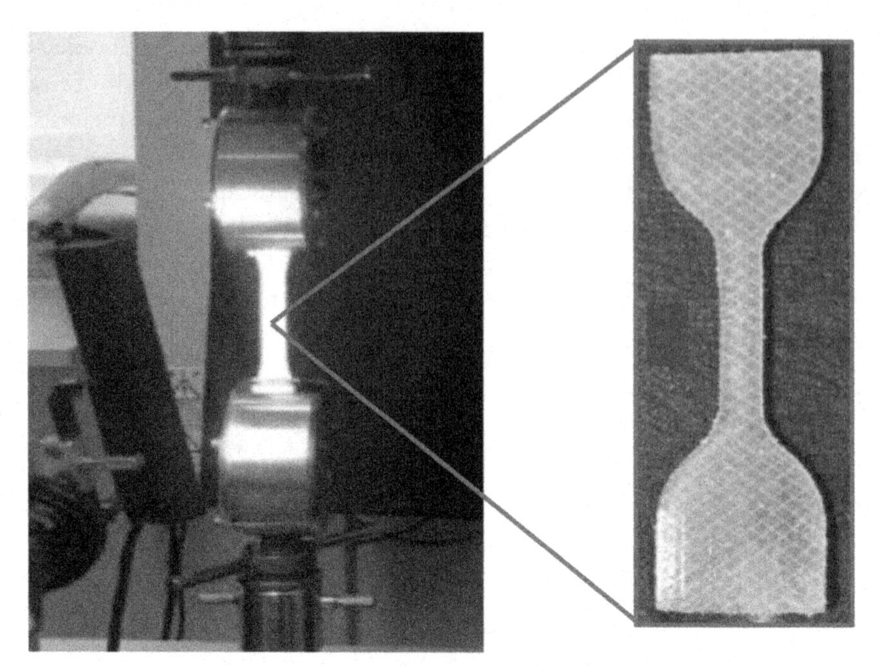

Figure 8.7 ***Typical Setup and Test Sample for Polymer Tensile Tests.***

Important results from a tensile test of polymer parts include the modulus, yield strength, ultimate tensile strength, and elongation.

Modulus is a measurement of the stiffness of the material. It defines the relationship between the stress and strain in an elastic material using the following equation:

$$E = \frac{\sigma}{\varepsilon}$$

where E is the modulus, σ is the tensile stress and ε is the corresponding tensile strain. Tensile stress and strain can be obtained using the following equations:

$$\sigma = \frac{F}{A}$$

$$\varepsilon = \frac{\Delta L}{L}$$

where F is the tensile force applied, A is the cross sectional area of the sample, ΔL is the change in length in the loading direction and L is the original length of the sample [20].

Yield strength of a material is defined as the stress at which the material starts to deform plastically. Ultimate tensile strength (UTS) is the capacity of the material to withstand loads and elongation is a measure of the ductility of a material. Elongation is the amount of strain a material can withstand before failure during a tensile test.

Typical parameters from polymer tensile tests are shown in Table 8.1.

3.1.2 Compressive Properties Measurement Methods

Compressive testing shows how the material will react when it is being compressed. Compression testing is able to determine the material's behavior or response under crushing loads and to measure the plastic flow behavior and ductile fracture limits of a material. Compression tests are important to measure the elastic and compressive fracture properties of brittle materials or low-ductility materials. Compression tests are also used to determine the modulus of elasticity, proportional limit, compressive yield point, compressive yield strength, and compressive strength. These properties are important to determine if the material is suited for specific applications or if it will fail under specific stresses. Compression tests can be carried out on polymer AM parts using the following standards:

ASTM D695 Standard test method for compressive properties of rigid plastics.

ISO 604 Plastics—determination of compressive properties, provide standards for testing of compressive properties.

The test sample should be a right cylinder with length twice its diameter or a right prism with length twice its principal width. Fig. 8.8 shows a polymer test coupon in compression test.

Typical parameters reported from compression tests are shown in Table 8.2.

3.1.3 Flexural Properties Measurement Methods

Flexural tests measure material's behavior subjected to simple beam loading. Flexural test is used to determine the material ductility by measuring the flexural strength and flexural modulus of polymers. Flexural modulus indicates the material stiffness when it is being flexed while flexural strength is related to the ability of the material to resist deformation under bending load. These characteristics are useful to determine if a material would begin to fracture or completely fracture under certain stresses or bending forces. This may lead to disastrous failure of the material when used in any application.

Table 8.1 Tensile properties of polymer

Process and material	Test method	E (MPa)	Yield strength (MPa)	UTS (MPa)	Elongation at yield (%)	Elongation at break (%)	References
FDM ABS (ABSplus-P430)	ASTM D638	2200	31	33	2	6	[21]
Polyjet VeroWhitePlus (RGD835)	ASTM D638	2000–3000	N.A.	50–65	N.A.	10–25	[22]

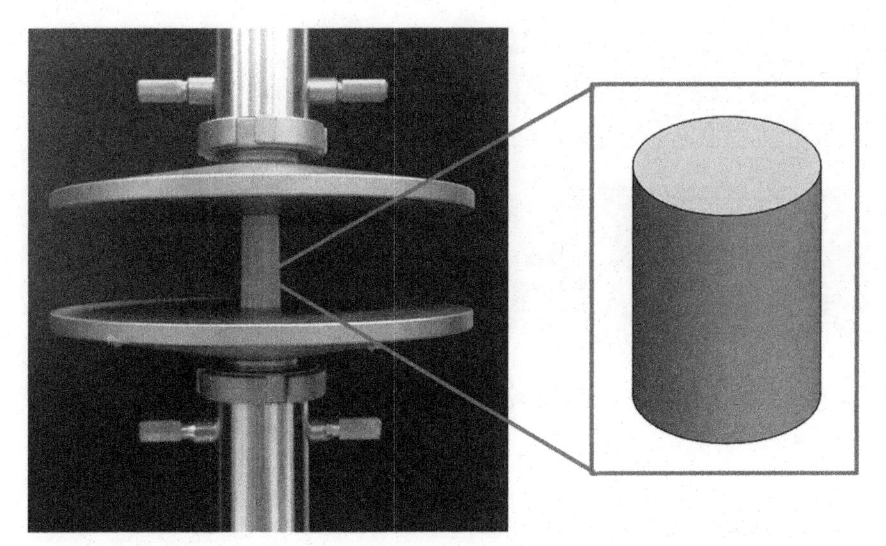

Figure 8.8 *Typical Setup for Polymer Compression Testing and Test Sample.*

Table 8.2 Compression properties of polymer

Process	Material	E (MPa)	Yield strength (MPa)	UTS (MPa)	References
FDM, 100% infill, extrusion temperature 340°C	PEEK	2016	71.15	83.61	[23]
FDM, 100% infill, extrusion temperature 100°C	Polywax	N.A.	N.A.	18–20	[24]

Flexural tests can be carried out using the following standards:

ASTM D790 Standard test methods for flexural properties of unreinforced and reinforced plastics and electrical insulating materials.

ISO 178-Plastics—determination of flexural properties.

The test sample is rectangular and should conform to ASTM D5947 Standard test methods for physical dimensions of solid plastics specimens.

Fig. 8.9 shows the three-point bending test on a plastic FDM coupon.

Typical parameters reported flexural tests shown on a commercially available material data sheet are shown in Table 8.3.

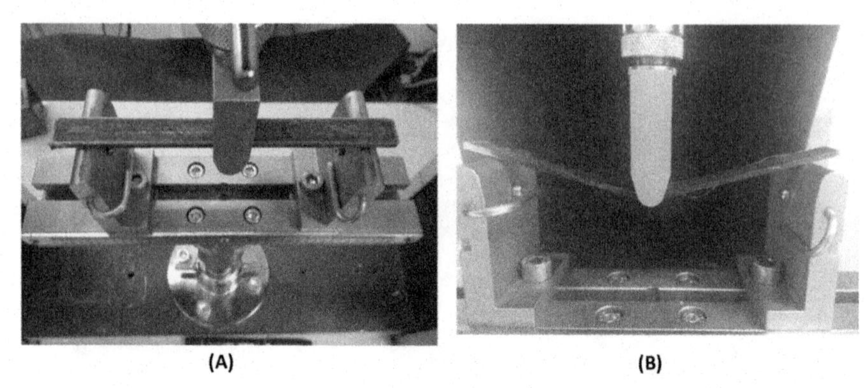

(A) (B)

Figure 8.9 (A) Top View of Three Point Bending Test (B) Front View of Three Point Bending Test.

Table 8.3 Flexural properties of polymer

Process	Material	Test method	Flexural Modulus (MPa)	Flexural strength (MPa)	Flexural strain at break (%)	References
FDM *XZ* axis	ABS (ABSplus-P430)	ASTM D790	2100	58	2	[21]
PolyJet	VeroWhitePlus (RGD835)	ASTM D790	2200–3000	75–110	N.A.	[22]

3.1.4 Impact Strength Measurement Methods

Impact testing is used to determine the ability of a material to withstand or absorb energy when experiencing a sudden high force or shock. The energy may be used to determine the toughness, fracture resistance, impact resistance and impact strength of a material depending on the test performed. These are important for the selection of material that will be used in high impact loading applications.

Impact tests can be carried out using the following standards:

ASTM D6110 Standard test method for determining the Charpy impact resistance of notched specimens of plastics.

ISO 179 Plastics—determination of Charpy impact properties.

Both provide standards for Charpy impact test, however, ISO 179 have many differences in testing parameters in comparison to ASTM D6110. The test gives results of the amount of energy to break the test specimen. Various specimen geometries and preparation considerations are specified by the standards.

Table 8.4 Impact properties of polymer

Process	Material	Test method	Izod notched impact (J/m)	References
FDM	ABS (ABSplus-P430)	ASTM D256	106	[21]
PolyJet	VeroWhitePlus (RGD835)	ASTM D256	20–30	[22]

ASTM D256 Standard test methods for determining the Izod pendulum impact resistance of plastics.

ISO 180 Plastics—determination of Izod impact strength.

Both provide standards for Izod impact test. It is similar to Charpy impact test. However, it has some parameters that are different such as the notch geometry, location and direction.

Typical parameters reported from impact tests shown on a commercially available material data sheet are shown in Table 8.4.

3.1.5 Other Properties

For liquid-based materials that may expand after production due to absorption of moisture in the environment, water absorption test would need to be done.

ASTM D570 Standard test method for water absorption of plastics.

ISO 62 Plastics—determination of water absorption.

Both provide standards for water absorption test which measures the rate at which the plastic absorbs water.

Density can help determine properties like strength to weight ratio and having near 100% density also means fewer pores.

ASTM D792 Standard test methods for density and specific gravity (relative density) of plastics by displacement.

This standard provides a method for measuring density using displacement of liquid at 23°C.

ISO 1183 Plastics—methods for determining the density of non cellular plastics.

This is similar to ASTM D792 but provides additional test method at 27°C.

Typical parameters reported for PolyJet VeroWhitePlus on a commercially available material data sheet is shown in Table 8.5.

Table 8.5 Other properties of VeroWhitePlus

Properties	Test method	Values	References
Water absorption	ASTM D570	1.1%–1.5%	[22]
Polymerized density	ASTM D792	1.17–1.18 g/cm^3	[22]

3.2 Metals

Metals similarly require testing of mechanical properties. Microstructure of the metal is often checked together with mechanical testing, as the type and size of the microstructure relates directly to its mechanical properties. ASTM has released the ASTM F3122 for evaluating mechanical properties of metal parts built via AM. The following sections will briefly explain some commonly tested properties.

3.2.1 Tensile Properties Measurement Methods

Tensile properties such as yield strength and ultimate tensile strength can be determined using the following standards:

ASTM E8/8M Standard test methods for tension testing of metallic materials.

ASTM E21 Standard test methods for elevated temperature tension tests of metallic materials.

ISO 6892 Metallic materials—tensile testing.

Different coupon types are available, including flat and cylindrical coupons. Different loading conditions, such as strain rates are also specified in the standards.

A typical tensile test set up and test specimen are shown in Fig. 8.10.

Typical parameters reported from tensile tests for metals are shown in Table 8.6.

The range of mechanical properties varies as the test results are dependent on their intended applications, as well as the processing conditions [27]. The processing conditions, such as thermal history have effect on the microstructure of the materials, which in turn, leads to range of results obtained [28,29].

3.2.2 Compressive Properties Measurement Methods

The compressive properties of metals can be measured using the following standards:

ASTM E9 Standard test methods of compression testing of metallic materials at room temperature.

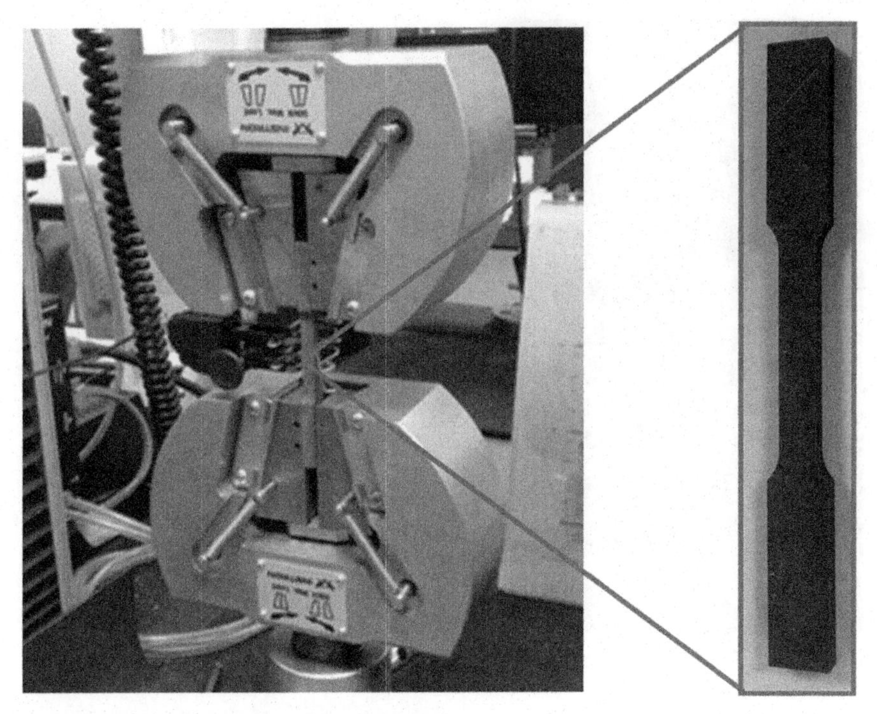

Figure 8.10 ***Tensile Test for a Metallic Sample.***

Table 8.6 Results from tensile test of metallic samples

Process	Material	Sample orientation	Yield stress (MPa)	UTS (MPa)	Elongation (%)	References
Laser Melting	Ti6Al4V	XY	1093 ± 64	1279 ± 13	6 ± 0.7	[25]
EBM	Ti6Al4V	XZ	950	1050	14	[26]

ASTM E209 Standard practice for compression tests of metallic materials at elevated temperatures with conventional or rapid heating rates and strain rates.

These provide standards for testing compressive strength at room temperature and elevated temperature, respectively. The stress-strain curve, compressive strength, and elastic modulus can be found.

A typical compression test set up and test sample are shown in Fig. 8.11.

Typical parameters reported from compression tests for metals are shown in Table 8.7.

Figure 8.11 *Compression Test for Metallic Sample.*

Table 8.7 Results from compression test of metallic samples

Process	Material	Porosity (%)	E (GPa)	Compressive strength (MPa)	References
SLM	Titanium	55	0.687	N.A.	[7]
SLM	Ti6Al4V	70	5.1 ± 0.3	155 ± 7	[24]

3.2.3 Hardness Measurement Methods

Hardness is a measure of how resistant the metal is to various kinds of permanent changes in shape when a compressive force is applied. The following standards detail methods that can be used to measure hardness of metals:

ASTM E10 Standard test method for Brinell hardness of metallic materials.

ISO 6506 Metallic materials—Brinell hardness test.

These provide standards for Brinell hardness test method. The test consists of using a sphere to indent a hole on the test coupon surface and the diameter of the indentation is used to calculate the hardness value.

ASTM E384 Standard test method for microindentation hardness of materials.

ISO 4545 Metallic materials—Knoop hardness test.

Table 8.8 Results from hardness test of metallic samples

Process	Material	Microhardness (HV)	Reference
SLM	Ti6Al4V	479–613	[29]
EBM	Ti6Al4V	358–387	[29]

These provide standards for Knoop and Vickers hardness test method, which is similar to Brinell hardness test but the indenter, is pyramid shape and the measure distance is the diagonal length of the indentation. The face angle of the pyramid is different between Knoop and Vickers test.

ASTM E18 Standard test methods for Rockwell hardness of metallic materials.

This provides standard for Rockwell hardness test which can use either pyramid or spherical indenters. Rockwell hardness test, however, does indentation multiple times at a single spot with increasing force. The depth of the indenter is measured by the machine thus directly giving the indentation depth.

Typical parameters reported from hardness tests for metals are shown in Table 8.8.

3.2.4 Fatigue Measurement Methods

Fatigue is the weakening of a material due to repeated applied loads or cyclic loading. Little work has been published for AM metals on the properties of fatigue. The standards that can be used to measure fatigue of metals are:

ASTM E466 Standard practice for conducting force controlled constant amplitude axial fatigue tests of metallic materials.

ISO 1099 Metallic materials—fatigue testing—axial force-controlled method.

These provide standards for axial force fatigue testing. The test involves pulling a specimen axially with a periodic force function which is often sinusoidal.

ASTM 647 Standard test method for measurement of fatigue crack growth rates.

ISO 12108 Metallic materials—fatigue testing—fatigue crack growth method.

These provide standards for notched samples to see how the material resists crack growth.

ASTM E2714 Standard test method for creep-fatigue testing.

Table 8.9 Results from fatigue test of metallic samples

Process	Material	Fatigue strength (MPa)	Fatigue life (Cycles)	References
EBM	Ti6Al4V	441	N.A.	[30]
EBM	Ti6Al4V	N.A.	28961 ± 5557	[31]

This provides the standard for the creep-fatigue test method. The test is similar to a fatigue test but at elevated temperature, so that it is able to give the strain/stress over time curve and stress-strain hysteresis curves.

Typical parameters reported from fatigue tests for metals are shown in Table 8.9.

3.2.5 Fracture Toughness Measurement Methods

Fracture toughness describes the ability of a material containing a crack to resist fracture. The following standards can be used to measure fracture toughness of metals:

ASTM E399 Standard test method for linear-elastic plane-strain fracture toughness K_{IC} of metallic materials.

ISO 12737 Metallic materials—determination of plane-strain fracture toughness.

ASTM E1820 Standard test method for measurement of fracture toughness.

ISO 12135 Metallic materials—unified method of test for the determination of quasistatic fracture toughness.

They provide standards for the measurement of the fracture toughness. ASTM E399 measures K_{IC} and ASTM E1820 gives the fracture toughness from *R*-curves.

Typical parameters reported from fracture toughness tests for metals are shown in Table 8.10.

3.2.6 Other Properties

There are some standards for consideration when characterizing metals, and they are:

ASTM E292 Standard test methods for conducting time-for-rupture notch tension tests of materials.

Table 8.10 Results from fracture toughness test of metallic samples

Process	Material	Orientation	K_{IC} (MPa/m)	References
EBM	Ti6Al4V	XZ	78.1 ± 2.3	[30]
EBM	Ti6Al4V	XY	96.9 ± 0.99	[30]

This provides the standards to find the rupture strength. The test finds the time taken to rupture a notched specimen which could be used to calculate the rupture strength to compare with that of a smooth specimen.

ASTM E111 Standard test method for Young's modulus, tangent modulus, and chord modulus.

This provides standard to find Young's modulus from tension and compression testing using ASTM E8 and ASTM E9, respectively.

ASTM E132 Standard test method for Poisson's ratio at room temperature.

This provides standard for finding the Poisson's ratio which finds the traverse and axial with a given load to calculate the Poisson's ratio.

ASTM E143 Standard test method for shear modulus at room temperature.

This provides standard for finding the shear modulus by torsion. The sample has to be cylindrical or a tube and the test gives the shear stress-strain curve which can be used to find the shear modulus.

3.2.7 Challenges

Due to the nature of the AM process, AM parts are often highly anisotropic. The standards listed previously are made for conventional manufacturing processes like casting, injection molding and extrusion, thus caution has to be taken to ensure the testing has taken anisotropy and build direction into account.

For plastics, suppliers that provide mechanical properties often give values in one build direction and build parameters [32,33]. However, mechanical properties vary with build direction and build parameters like hatching distance [34,35]. Fig. 8.12 shows different tensile coupons printed in different build directions.

Metal additive manufacturing parts are often made from metal powders and suppliers of metal powders do not provide mechanical properties of the material but powder properties instead. Metal AM parts have their properties benchmarked against bulk material properties.

Build direction is not the only factor influencing mechanical properties. Build parameters like scanning strategy, hatching distance, and temperature can interact together and affect final part properties. Mohamed et al. used neural networks to optimize build parameters in FDM for dynamic mechanical analysis, which involves several inputs like hatch spacing, track width, and print direction [36]. In order to optimize a property, several build

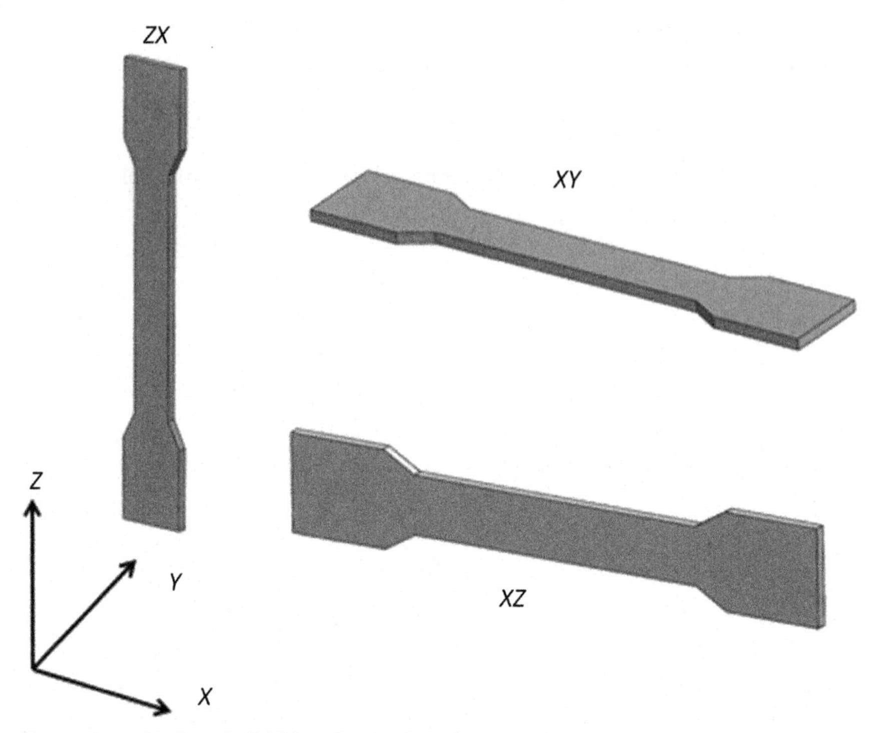

Figure 8.12 ***Various Build Direction for Tensile Coupon.***

parameters have to be considered together, making it complex to classify and test AM material mechanical properties. Ahn et al. showed how different hatching designs in FDM can help certain applications like snap fit clips [37]. However in testing for tensile strength alone, it may not be enough to encompass actual application situations where parts have complex geometry. Due to the nature of AM parts, using testing and analysis methods for fiber-reinforced polymers can help in better characterizing FDM parts [38].

Metal parts are also similarly affected by interacting build parameters. Ning et al. found that not only the raster angle affects tensile strength, length of the hatch lines causes difference in shrinkage, leading to poorer surface finish and differing geometry which affects mechanical properties for SLM [35]. Siddique et al. tested for the tensile strength of AlSi12 coupons made with different energy density, base plate heating and heat treatment, and all the different combinations gave different ultimate tensile strength and elongation [39]. Similar results were seen in the fatigue life of AlSi10 samples, where build direction, build plate temperature, and heat treatment affected the fatigue life cycle [40]. Classifying sample preparation has to be clearly defined, as heat treated and as built coupons can lead to greatly varying mechanical properties.

3.3 Potential of Computational Methods

Computational models can help to reduce time and materials spent on experiment [41]. However, the physics of the materials and process have to be well understood in order to have accurate models. Bellini and Guceri used isotropic and anisotropic properties in finite element analysis (FEA) to evaluate stresses; directional tensile strengths of FDM ABS parts were found from experiment and used in the model [42]. The anisotropic model gave different stresses compared to the isotropic model. FEM models should consider the anisotropic properties of the material to model deformation and stresses. Gorski et al. modeled the CAD model with the raster angle and air gap in order to better evaluate FDM ABS mechanical properties. The model was accurate in getting the bending stress. However, internal stresses were still inaccurate [43]. Other mechanical properties like compressive stress have been modeled. However, it also has the limitation of being only accurate for linear analysis and cannot take into account nonlinear situations like buckling [44]. FEA is not the only computational method, Vijayaraghavan et al. used the neural network to predict wear strength based on various build parameters [45]. However, it should be noted that such computational method requires extensive amount of data to be accurate, which needs a considerable number of experiments, costing time and money. Residual stress of a single layer of Inconel was evaluated using FEA, predicting the deformation after forming one layer using SLM [46]. However, such simulation requires significant computational power and time. Simulation on an actual part can take years. In order to reduce simulation time, Zheng et al. wrote their own simulation software that has a moving fine mesh where the laser heat is while having coarser mesh everywhere else. This reduced computational time by considerable amounts. They have commercialized the software as 3DSIM (3DSIM LLC) [47,48].

4 BENCHMARKING OF LOW-COST PRINTERS

3D printers that once cost more than $10,000 are now available for a fraction of that price. The low-cost 3D printers available in the market typically are the FDM and Stereolithography (SLA) or digital light processing (DLP) printers. FDM 3D printers are generally more affordable than SLA or DLP printers. A low-cost FDM 3D printer could be obtained easily from $200 to $2000, while a low-cost SLA or DLP printer usually costs above $3000. The FDM printers are commonly used by home users and hobbyists due to its ease of operation and relatively safer and simpler material handling. Nevertheless, the SLA and DLP remains popular among the small businesses and professionals, such as jewellery makers and dentists who require higher resolution and smoother parts.

The proliferation of low-cost 3D printers, including both open source and closed source printers actively pushes for strong growth in both consumer and professional markets. Though many of the low-cost printers offer the same functionality, each 3D printer has different features such as printer design, printing performance, print quality, as well as material consumption and waste. Benchmarking is thus an important and useful tool to evaluate these 3D printers' quality and capability in a comparative manner [49]. Besides the cost of the 3D printers which can be easily compared quantitatively, there are many others factors that should be considered in the benchmarking process for the low-cost 3D printers, and they will be discussed in the following sections.

4.1 Print Quality

The print quality is one of the main performance indicators for every 3D printer. Similar to commercial professional 3D printers, the performance of the low-cost printers can be quantitatively evaluated and compared by printing the same benchmark model using different 3D printers [4]. Benchmark test artifacts including various features, such as thin walls, rectangular and cylindrical bosses, through and blind holes, inclines, and notches, can be specially designed to assess the print quality in terms of the surface quality of the printed surfaces, dimensional accuracy and tolerances, repeatability, and other geometry limitations [2,50–52]. Benchmarking also allows the users to identify the parameters tested, for instance, layer thickness, raster width, nozzle speed, and so on, that give the lowest dimensional accuracy discrepancies for the benchmarking model, therefore identifying the best achievable practices and processes [4,53].

4.2 Build Time and Build Volume

The build envelope size determines the maximum object size that can be printed and the number of parts that the printer is able to build at a time. Although a low-cost printer is compact and has a desktop size, typically with a build volume smaller than $150 \times 150 \times 150$ mm^3, some printers offer much bigger build volume at an affordable price. In the case of the object that is too large to be fit into the build envelope, the object has to be printed in pieces for assembly later. Studies have shown that building multiple parts is generally more efficient than printing a single part for various systems through calculating the build time scaling factor [49].

The build speed in FDM is dependent on several printing parameters, including the movement speed, nozzle size and layer height. It is often a trade-off between build speed and resolution in 3D printing processes. In order to speed up the printing, the surface and finish quality would have to be sacrificed by increasing layer thickness and nozzle diameter, as well as by reducing fill density.

4.3 Material Usage and Waste

In addition to the material used to build the intended model, one should take into account of the amount of waste material when calculating the total material usage. The waste materials include the support material that supports overhangs, of which the quantity could be influenced by the part orientation and other settings of the printing, and any model material (resin or filament) used in printing that does not end up in the finished part, but is disposed of during or after the printing process. The model material waste is generally small in amount and it can include the filament extruded at the beginning of every print, additional raft, skirt and brim to enhance bed adhesion, stabilization, as well as filament extrusion.

Research has shown that the material use for printing multiple parts simultaneously in one build is approximately the same as for printing only one part at a time using FDM and SLA processes [49]. The percentage of waste generated is generally consistent across different models of FDM machines [49]. The exact amount of waste generated is dependent on the part geometry and the support material generated through the software.

4.4 Safety

FDM printer has many moving mechanical parts, including motors, gears, and belts. The extrusion tip would be heated to temperature as high as 220°C to melt the filaments while the printer is in operation. Some FDM printers also include a heated build plate to enhance part adhesion to the platform. All these could potentially lead to pinching and entrapment hazards, as well as burn hazard. Although open-framed printers offer visibility of the print job and easy access to the build platform and extruder, fully enclosed printers are safer by preventing accessibility to these moving and heated components and thus minimizing the risk of injury. The enclosure

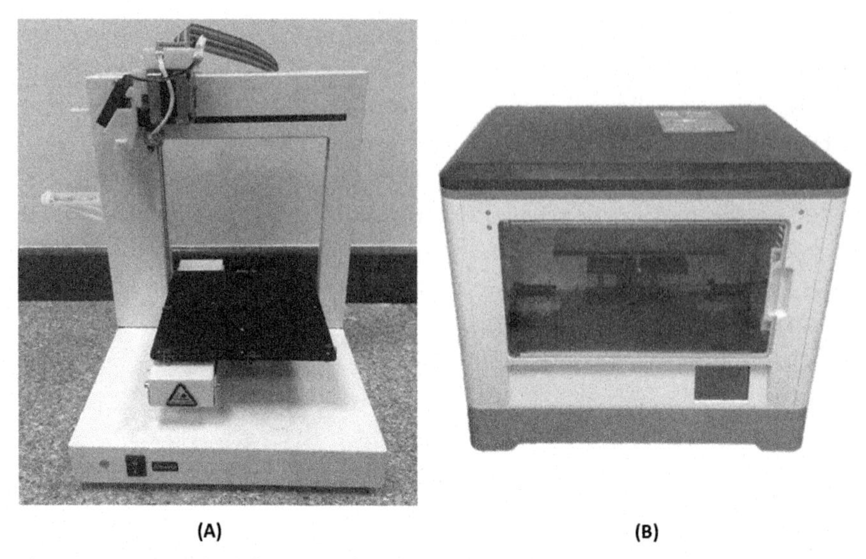

(A) (B)

Figure 8.13 Examples of (A) open framed and (B) fully enclosed FDM printers.

also reduces noise and possible odor when printing with ABS. Examples of open-framed printers and fully enclosed printers are shown in Fig. 8.13.

Studies done by Illinois Institute of Technology has discovered that potentially harmful ultrafine particles (UFP) were emitted from the operation of desktop 3D printers using the PLA and ABS filaments [54,55]. These UFPs are the thermal decomposition products from ABS and PLA. The results have shown that high emission rate of about 20 billion particles per minute for PLA feedstock and about 200 billion particles per minute for ABS. Studies have shown that elevated concentration of UFPs is associated with adverse health effects, including cardio–respiratory mortality, stroke and asthma symptoms [56,57]. However, most FDM printers, including the industrial grade printers, are not equipped with any exhaust ventilation or filtration accessories. Results herein suggest operating FDM printers in an inadequately ventilated environment could be hazardous. This is further supported by tests showing ABS thermal decomposition at high temperature emits carbon monoxide and hydrogen cyanide and exposure to ABS thermal decomposition products display toxic effect in rats and mice [58–62].

Resin-based SLA printers require the use of resin material and solvents for postprocessing. The resin and solvents may pose severe safety hazard and environmental hazard if they are not handled and disposed of properly. Safety precautions, for instance, wearing gloves and safety glasses, must be abided to prevent skin contact with the resin. It is also essential to handle the waste resin and solvent disposal responsibly. In addition, ventilation is crucial for operating the SLA because of the volatile toxic organic compounds used as the monomers and photoinitiators. Besides safety concern over the uncured resin, study also has shed light on the toxicity levels of the SLA and FDM printed parts. It is discovered that the printed parts were considerably toxic to zebrafish embryos, with the SLA printed parts significantly more toxic than the FDM printed parts [63].

4.5 Machine Design and Ease of Use

The printer has to be sufficiently heavy in order to maintain stability in its operational and neutral position. The printer has to remain steady to withstand rigorous printing movements and unintended substantial external force during the printing process, so that the print is more precise and more reliable. For the frame or chassis, aluminum or steel is often preferred over lightweight plastics for stability.

Many of the low-cost 3D FDM printers are equipped with heated platform. Heated platform is essential in printing ABS which is printed at higher temperature because ABS would warp due to the rapid and uneven rate of cooling. ABS adheres better to the heated platform and the ABS printed parts can be cooled slowly without shrinking to reduce warping.

In addition, since most low-cost 3D printers are targeted at the home-based users, the machine should be easy to assemble, easy to use, and easy to maintain. The learning curve for low-cost printer is usually less steep and thus on-site training is not usually provided. Other value-added features that a low-cost printer could offer include automatic bed leveling, on-board control panel, material compatibility, as well as universal filament compatibility.

4.6 Portability and Connectivity

Portability is another factor that should be considered to benchmark among the low-cost printers. On top of weight and size concerns, the connectivity and compatibility of whether the machine allows printing without having a computer connected via USB cable to the 3D printer for the entire duration of the printing process, should be evaluated to determine its portability. If a computer is required to be connected to the printer at all times, it would be difficult to transport the entire system for use in another location [49]. Today, many low-cost 3D printers have the ability to print a part by reading the .stl files directly from SD card or USB drive while few low-cost 3D printers offer wireless connectivity, which allows users to send files wirelessly to the 3D printer via 802.11 Wi-Fi or a direct, peer-to-peer link.

4.7 Relevant Standards and Guidelines

There are a number of existing safety standards that are applicable to equipment associated with 3D printers and could adequately cover the safety of 3D printers. As the energy sources and safeguards for 3D printers are similar to those in other more complex forms of IT equipment and office appliances certified to IEC TC108 standards, such as 2D laser and inkjet printers, it is possible to apply these available standards directly on 3D printers. 3D printer typically falls under two main categories, electronic products and industrial machinery, each with its own safety standards. A low-cost 3D printer that is used in homes, schools, offices, and laboratories for small-scale production can be covered under nonindustrial requirements. Despite the exponential growth of this relatively new consumer technology, it is not yet a requirement for 3D printers to meet specific product safety standards. Only a handful of the commercial low-cost 3D printers have undergone certification and meets the requirement to be certified safe for home use or even children use.

The following standards can be considered applicable for low-cost 3D printers:

IEC 62368-1 Audio/video, information and communication technology equipment—Part 1: Safety requirements.

This standard describes the safety of electrical and electronic equipment with a rated voltage not exceeding 600 V. There are six major hazards described in this standard including electrically-caused pain or injury, electrically-caused fire and injury caused by hazardous substances, mechanically-caused injury and thermally-caused injury. It contains the tests and their compliance criteria, as well as the requirements to safeguard against different potential hazard sources. This is one of the most common standards used for certification of 3D printers.

IEC 60950-1 Information Technology Equipment—Safety—Part 1: General requirements.

This standard intends to reduce the risk of injury or damage to the users and service personnel from encountering seven major hazards: electric shock, energy and heat related hazards, fire, mechanical, radiation, and chemical hazards. It contains potential hazard elements and possible preventive measures. This is one of the most common standards used for certification of 3D printers.

IEC TR 62471-2 Photobiological safety of lamps and lamp systems—Part 2: guidance on manufacturing requirements relating to nonlaser optical radiation safety.

This report serves as a guide for optical radiation safety assessment and for allocation of safety measures of lamps and other broadband light sources including LEDs and lamps with projection systems that are used in the DLP printers, as well as the UV oven for postcuring.

IEC 60227 Polyvinyl chloride insulated cables of rated voltages up to and including 450/750 V.

This standard contains the requirement and test methods for rigid and flexible cables with insulation, and sheath, based on polyvinyl chloride, that is commonly found in every printers' cables.

IPC-1331 Voluntary safety standard for electrically heated process equipment.

This voluntary standard establishes minimum requirements to the design, installation, operation, and maintenance of electrically heated process equipment in order to minimize potential electrical hazards. This is applicable in the FDM printers for heating the platform and the nozzle.

ISO 13732-1 Ergonomics of the thermal environment—Methods for the assessment of human responses to contact with surfaces—Part 1: Hot surfaces

This standard provides guidance to specify temperature limit values for hot surfaces and provides temperature threshold values for burns that occur when human skin is in contact with a hot solid surface.

5 QUESTIONS

1. State the key considerations in designing a test artifact and explain how they are important for benchmarking.
2. Describe the functions of benchmarking.
3. Explain the terms "roundness" and "straightness", and briefly describe how they can be measured.
4. State the tensile properties that can be obtained from a tensile test.
5. Describe the differences between tensile and compression tests.
6. List the key considerations when benchmarking a low-cost 3D printer.

REFERENCES

[1] C.K. Chua, K.F. Leong, 3D Printing and Additive Manufacturing: Principles and Applications, fifth ed., World Scientific Publishing Company, Singapore, (2017).
[2] H. Yang, C.J. Lim, Y. Liu, X. Qi, Y.L. Yap, V. Dikshit, et al. Performance evaluation of Projet multi-material jetting 3D printer, Virtual Phys. Prototyp 12 (2017) 95–103.
[3] U. Berger, Aspects of accuracy and precision in the additive manufacturing of plastic gears, Virtual Phys. Prototyp. 10 (2015) 49–57.

[4] F.A. Cruz Sanchez, H. Boudaoud, L. Muller, M. Camargo, Towards a standard experimental protocol for open source additive manufacturing, Virtual Phys. Prototyp. 9 (2014) 151–167.
[5] N. Hopkinson, T.B. Sercombe, Process repeatability and sources of error in indirect SLS of aluminium, Rapid Prototyp. J. 14 (2008) 108–113.
[6] S.L. Campanelli, G. Cardano, R. Giannoccaro, A.D. Ludovico, E.L.J. Bohez, Statistical analysis of the stereolithographic process to improve accuracy, Comput. Aided Design 39 (2007) 80–86.
[7] S.L. Sing, W.Y. Yeong, F.E. Wiria, B.Y. Tay, Characterization of titanium lattice structures fabricated by selective laser melting using an adapted compressive test method, Exp. Mech. 56 (2016) 735–748.
[8] Y. L. Yap, C.C. Wang, H.K. J. Tan, V. Dikshit, and W. Y. Yeong, Benchmarking of Material Jetting Process: Process Capability Study, in: Proceedings of the 2nd International Conference on Progress in Additive Manufacturing, Singapore, 2016, pp. 513–518.
[9] T.H.C. Childs, N.P. Juster, Linear and geometric accuracies from layer manufacturing, CIRP Ann. Manufact. Technol. 43 (1994) 163–166.
[10] F. Xu, Y.S. Wong, H.T. Loh, Toward generic models for comparative evaluation and process selection in rapid prototyping and manufacturing, J. Manufact. Syst. 19 (2001) 283–296.
[11] M. Mahesh, Y.S. Wong, J.Y.H. Fuh, H.T. Loh, Benchmarking for comparative evaluation of RP systems and processes, Rapid Prototyp. J. 10 (2004) 123–135.
[12] M. Mahesh, Y.S. Wong, J.Y.H. Fuh, H.T. Loh, A six-sigma approach for benchmarking of RP&M prcoesses, Int. J. Adv. Manufact. Technol. 31 (2006) 374–387.
[13] K. Abdel Ghany, S.F. Moustafa, Comparison between the products of four RPM systems for metals, Rapid Prototyping J. 12 (2006) 86–94.
[14] J.M. Lee, M. Zhang, W.Y. Yeong, Characterization and evaluation of 3D printed microfluidic chip for cell processing, Microfluid. Nanofluid. 20 (2016) 5.
[15] J.A.S.S.P. Moylan, A.L. Cooke, K.K. Jurrens, M.A. Donmez, Proposal for a standardized test artifact for additive manufacturing machines and processes, in: the 23rd International Solid Free Form Symposium—An Additive Manufacturing Conference, Austin, TX, USA, 2012, pp. 902–920.
[16] J. Van Keulen, Density of porous solids, Matériaux et Construction 6 (1973) 181–183.
[17] Standard Test Methods for Density of Compacted or Sintered Powder Metallurgy (PM) Products Using Archimedes; Principle, ed: ASTM International, 2015.
[18] R.C. Hibbeler, Mechanics of materials, 4th ed., Pearson, United Kingdom, (2017).
[19] C.W. de Silva, Mechanics of materials, 4th ed., CRC Press, United States of America, (2013).
[20] A. International, ASM Handbook Volume 8 Mechanical Testing and Evaluation, 4th ed., 2015.
[21] (2015, 23 January 2017). ABSplus Spec sheet. Available from: http://usglobalimages.stratasys.com/Main/Files/Material_Spec_Sheets/MSS_FDM_ABSplusP430.pdf
[22] (2016, 23 January 2017). PolyJet materials data sheet. Available from: http://usglobalimages.stratasys.com/Main/Files/Material_Spec_Sheets/MSS_PJ_PJMaterialsDataSheet.pdf?v=635785205440671440
[23] K.M. Rahman, T. Letcher, R. Reese, "Mechanical Properties of Additively Manufactured PEEK Components Using Fused Filament Fabrication," in ASME 2015 International Mechanical Engineering Congress and Exposition, 2015, p. V02AT02A009.
[24] J. Wang, H. Xie, Z. Weng, T. Senthil, L. Wu, A novel approach to improve mechanical properties of parts fabricated by fused deposition modeling, Mater Design 105 (2016) 152–159.
[25] V. Cain, L. Thijs, J. Van Humbeeck, B. Van Hooreweder, R. Knutsen, Crack propagation and fracture toughness of Ti6Al4V alloy produced by selective laser melting, Add. Manufact. 5 (2015) 68–76.

[26] X. Zhao, S. Li, M. Zhang, Y. Liu, T.B. Sercombe, S. Wang, et al. Comparison of the microstructures and mechanical properties ofTi–6Al–4V fabricated by selective laser melting and electron beam melting, Mater. Design 95 (2016) 21–31.
[27] J.J. Lewandowski, M. Seifi, Metal additive manufacturing: a review of mechanical properties, Ann. Rev. Mater. Res. 46 (2016) 151–186.
[28] Y. Kok, X. Tan, S.B. Tor, C.K. Chua, Fabrication and microstructural characterisation of additive manufactured Ti-6Al-4V parts by electron beam melting, Virtual Phys. Prototyp. 10 (2015) 13–21.
[29] S.L. Sing, J. An, W.Y. Yeong, F.E. Wiria, Laser and electron-beam powder-bed additive manufacturing of metallic implants: a review on processes, materials and designs, J. Ortho. Res. 34 (2016) 369–385.
[30] W.E. Frazier, Metal additive manufacturing: a review, J. Mater. Eng. Perform. 23 (2014) 1917–1928.
[31] K.S. Chan, M. Koike, R.L. Mason, T. Okabe, Fatigue life of titanium alloys fabricated by additive layer manufacturing techniques for dental implants, Metal. Mater. Transact. A 44 (2013) 1010–1022.
[32] J. Kotlinski, Mechanical properties of commercial rapid prototyping materials, Rapid Prototyp. J. 20 (2014) 499–510.
[33] G. Kim, Y. Oh, A benchmark study on rapid prototyping processes and machines: quantitative comparisons of mechanical properties, accuracy, roughness, speed, and material cost, Proceedings of the Institution of Mechanical Engineers, Part B: Journal of Engineering Manufacture, vol. 222, pp. 201–215, 2008.
[34] O.S. Es-Said, J. Foyos, R. Noorani, M. Mendelson, R. Marloth, B.A. Pregger, Effect of Layer Orientation on Mechanical Properties of Rapid Prototyped Samples, Materials and Manufacturing Processes, vol. 15, pp. 107–122.
[35] Y. Ning, Y. Wong, J. Fuh, Effect and control of hatch length on material properties in the direct metal laser sintering process, Proceedings of the Institution of Mechanical Engineers, Part B: Journal of Engineering Manufacture, vol. 219, pp. 15–25, 2005.
[36] O.A. Mohamed, S.H. Masood, J.L. Bhowmik, Analytical modelling and optimization of the temperature-dependent dynamic mechanical properties of fused deposition fabricated parts made of PC-ABS, Materials 9 (2016) 895.
[37] S.-H. Ahn, M. Montero, D. Odell, S. Roundy, P.K. Wright, Anisotropic material properties of fused deposition modeling ABS, Rapid Prototyp. J. 8 (2002) 248–257.
[38] N. Hill, M. Haghi, Deposition direction-dependent failure criteria for fused deposition modeling polycarbonate, Rapid Prototyp. J. 20 (2014) 221–227.
[39] S. Siddique, M. Imran, E. Wycisk, C. Emmelmann, F. Walther, Influence of process-induced microstructure and imperfections on mechanical properties of AlSi12 processed by selective laser melting, J. Mater. Proc. Technol. 221 (2015) 205–213.
[40] E. Brandl, U. Heckenberger, V. Holzinger, D. Buchbinder, Additive manufactured AlSi10Mg samples using selective laser melting (SLM): microstructure, high cycle fatigue, and fracture behavior, Mater. Design 34 (2012) 159–169.
[41] S.L. Sing, Y. Miao, F.E. Wiria, W.Y. Yeong, Manufacturability and mechanical testing considerations of metallic scaffolds fabricated using selective laser melting: a review, Biomed. Sci. Eng. 2 (2016) 18–24.
[42] A. Bellini, S. Güçeri, Mechanical characterization of parts fabricated using fused deposition modeling, Rapid Prototy.p J. 9 (2003) 252–264.
[43] F. Górski, W. Kuczko, R. Wichniarek, A. Hamrol, Computation of mechanical properties of parts manufactured by fused deposition modeling using finite element method, in: 10th International Conference on Soft Computing Models in Industrial and Environmental Applications, 2015, pp. 403–413.
[44] L. Villalpando, H. Eiliat, R. Urbanic, An optimization approach for components built by fused deposition modeling with parametric internal structures, Procedia CIRP 17 (2014) 800–805.

[45] V.Vijayaraghavan, A. Garg, J.S.L. Lam, B. Panda, S. Mahapatra, Process characterisation of 3D-printed FDM components using improved evolutionary computational approach, Int. J Adv. Manufact. Technol. 78 (2015) 781–793.
[46] B. Cheng, S. Shrestha, K. Chou, Stress and deformation evaluations of scanning strategy effect in selective laser melting, Add. Manufactur. 12 (2016) 240–251.
[47] D. Pal, N. Patil, K. Zeng, B. Stucker, An integrated approach to additive manufacturing simulations using physics based, coupled multiscale process modeling, J. Manufact. Sci. Eng. 136 (2014) 061022.
[48] K. Zeng, D. Pal, H. Gong, N. Patil, B. Stucker, Comparison of 3DSIM thermal modelling of selective laser melting using new dynamic meshing method to ANSYS, Mater. Sci. Technol. 31 (2015) 945–956.
[49] D.A. Roberson, D. Espalin, R.B. Wicker, 3D printer selection: a decision-making evaluation and ranking model, Virtual Phys. Prototyp. 8 (2013) 201–212.
[50] W. M. Johnson, M. Rowell, B. Deason, M. Eubanks, Benchmarking evaluation of an open source fused deposition modeling additive manufacturing system, in: the 22nd Annual International Solid Freeform Fabrication Symposium—An Additive Manufacturing Conference, Austin, TX, USA, 2011, pp. 197–211.
[51] L. Yang, M.A. Anam, An investigation of standard test part design for additive manufacturing, in: The 25th Annual International Solid Freeform Fabrication Symposium—An Additive Manufacturing Conference, Austin, TX, USA, 2014, pp. 901–922.
[52] J.-Y. Lee, W.S. Tan, J. An, C.K. Chua, C.Y. Tang, A.G. Fane, et al. The potential to enhance membrane module design with 3D printing technology, J. Membr. Sci. 499 (2016) 480–490.
[53] L.M. Galantucci, I. Bodi, J. Kacani, F. Lavecchia, Analysis of dimensional performance for a 3D open-source printer based on fused deposition modeling technique, Procedia CIRP 28 (2015) 82–87.
[54] B. Stephens, P. Azimi, Z. El Orch, T. Ramos, Ultrafine particle emissions from desktop 3D printers, Atmos. Environ. 79 (2013) 334–339.
[55] P. Azimi, D. Zhao, C. Pouzet, N.E. Crain, B. Stephens, Emissions of ultrafine particles and volatile organic compounds from commercially available desktop three-dimensional printers with multiple filaments, Environ. Sci. Technol. 50 (2016) 1260–1268.
[56] M. Stolzel, S. Breitner, J. Cyrys, M. Pitz, G. Wolke, W. Kreyling, et al. Daily mortality and particulate matter in different size classes in Erfurt, Germany, J. Exposure Sci. Environ. Epidemiol. 17 (2007) 458–467.
[57] H.R.P.o.U. Particles, Understanding the health effects of ambient ultrafine particles, Health Effects Institute, Boston, MA, 2013.
[58] Y. Deng, S.-J. Cao, A. Chen, Y. Guo, The impact of manufacturing parameters on submicron particle emissions from a desktop 3D printer in the perspective of emission reduction, Build. Environ. 104 (2016) 311–319.
[59] J.V. Rutkowski, B.C. Levin, Acrylonitrile–butadiene–styrene copolymers (ABS): Pyrolysis and combustion products and their toxicity—a review of the literature, Fire Mater. 10 (1986) 93–105.
[60] T.L. Zontek, B.R. Ogle, J.T. Jankovic, S.M. Hollenbeck, An exposure assessment of desktop 3D printing, J. Chem. Health Safe. 24 (2016) 15–25.
[61] A. Zitting, H. Savolainen, Effects of single and repeated exposures to thermo-oxidative degradation products of poly(acrylonitrile-butadiene-styrene) (ABS) on rat lung, liver, kidney, and brain, Arch. Toxicol. 46 (1980) 295–304.
[62] M.M. Schaper, R.D. Thompson, K.A. Detwiler-Okabayashi, Respiratory responses of mice exposed to thermal decomposition products from polymers heated at and above workplace processing temperatures, Am. Ind. Hygiene Assoc. J. 55 (1994) 924–934.
[63] S.M. Oskui, G. Diamante, C. Liao, W. Shi, J. Gan, D. Schlenk, et al. Assessing and reducing the toxicity of 3d-printed parts, Environ. Sci. Technol. Lett. 3 (2016) 1–6.

CHAPTER NINE

Quality Management Framework in Additive Manufacturing

Contents

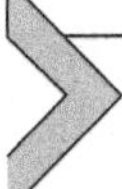

1 THE NEED FOR QUALITY MANAGEMENT FRAMEWORK IN ADDITIVE MANUFACTURING

Quality enables a user to characterize and determine which product is better than the other. The quality level of products and services indicates not only their intended function and performance but also their perceived value and benefit to the customer [1–4]. Despite the growing role of AM in the industry, an appropriate quality management framework has yet to be established to ensure quality and process consistency in manufacturing. To achieve quality is to effectively meet the customer needs. Thus, in an organization that is providing AM as a service, a quality management framework

Standards, Quality Control, and Measurement Sciences in 3D Printing and Additive Manufacturing
http://dx.doi.org/10.1016/B978-0-12-813489-4.00009-X

must be established to assure and improve the quality of the manufacturing, as well as the service. This can be achieved by focusing on continuous feedback on the existing process to refine and improve on the current practices. Therefore, in the AM industry, organizations are required to have a quality framework that additionally addresses the new concerns specific to AM, in addition to adopting and committing to the approaches and expectations defined in quality management standards, such as ISO 9001:2015.

International quality standards are published by ISO and other relevant standards bodies. Certification is, however, performed by the relevant national certification bodies, business management consultancy companies, and auditing companies. These auditing companies are usually accredited by a national body for auditing. To attain approval for an organization's quality management system (QMS), the organization has to be audited by an accredited consultant company. If the audit is satisfactory, the consultant puts up a recommendation for approval for that organization. AM has not yet reached a state of maturity whereby the existing QMS address the unique challenges that arise from the fundamentally different nature of the process. Existing AM operating companies should actively participate in the development of an AM QMS that addresses these challenges, providing necessary information and feedback. This enables operating companies to optimize their processes based on a priori knowledge and to assure a stable process with defined product quality requirements related to customer needs and expectations.

There are many forms of quality oriented programs introduced by different companies throughout the history of manufacturing. Lean manufacturing, total quality management, six-sigma, zero defects, etc., are some of the manufacturing methods commonly employed by companies to improve their workflow [4,5]. Such processes and workflows can be adopted for AM and be integrated into the QMS defined specially for AM industry.

Any manufacturer using AM would have to be aware of and compliant to the requirements imposed by both their customers and the regulating authorities. Organizations using AM as a service should adopt ISO 9001 to form the fundamental basis of proper documentation of the processes. Although this methodology does not guarantee product quality, it ensures that the process is performed in a quality manner. The ISO 9001:2015 QMS standard explains that the organization has to consider the following:

- External and internal issues.
- Requirements of relevant interested parties, who are customers, statutory, and regulatory requirements.
- The products and services of the organization.

To be certified, a typical AM organization can adopt the requirements stated in ISO 9001:2015 [6] and its requirements will be later discussed in Section 1.1 to 1.8.

ISO QMS employs the concept of a manufacturing environment framework with the involvement of all relevant stakeholders. The quality management framework defined in ISO 9001:2015 consists of seven sections, which are [6]:

- Context of the organization.
- Leadership and commitment.
- Planning.
- Support.
- Operation.
- Performance evaluation.
- Improvement.

The framework assists organizations in developing the proposed QMS through effective application of the system, process improvement, and assurance.

1.1 Context of the Organization

The context of an organization consists of both internal and external factors which influence its direction and its ability to attain certification for QMS. The organization has to monitor both internal and external factors and act on items that are flagged for action. Some examples of external factors are market of interest, environment of a country, technological limitations, and so on. Internal factors may include values, culture, and knowledge of an organization. Taking Singapore as an example, the lack of land space and natural resources makes large scale manufacturing infeasible. However, AM would be more applicable in this country for small-scale customized fabrication. Furthermore, Singapore houses the Singapore Centre for 3D Printing which is one of the world's largest AM research center in an institute of higher learning. Through collaborations, companies which intend to invest in AM in Singapore can reap benefits from the center in terms of knowledge and technical know-how.

There are many uncertainties in AM processes, and the organization has to know how these uncertainties affect them. The organization has to meet their customers' requirements, as well as those defined by the statutory and regulatory agencies, which are currently not well established. An option for an AM organization is to develop benchmarks measuring the performance of raw materials, process guidelines, postprocessing, material measurement, and material testing. From these benchmarks, they are able to define the minimum requirements needed to achieve a good quality printed part. With the development of a QMS, they also must consistently monitor, review, and improve their process as part of the QMS. The organization could also collaborate with their relevant country's standards bodies to work on new requirements for AM, and define best practices.

1.2 Leadership and Commitment

The top management of an AM organization plays a crucial role in its commitment toward quality of products and services. The management has to take the lead in ensuring the accountability of implementation, usefulness, and impact of the QMS. For example, organizations that are engaged in the AM service bureau business have to ensure that the additively manufactured parts are delivered on time to their customers, assisted by the proper implementation of policies and objectives of the QMS. They also have to ensure that the policies and objectives are in alignment with both the strategic and business direction of the organization.

To implement QMS in an effective manner, the management has to ensure that sufficient training and resources are allocated for the staff to carry out their tasks. In addition, prior knowledge of AM machines, processes, and materials is required to deliver quality work. Thus, they need to emphasize continuous improvement using process-based approaches and risk-based thinking, and communicate this information to all relevant stakeholders in the organization [6].

The top management also has to engage the people on the ground to contribute to the QMS and support the different intermediate management roles in the organization to ensure that they portray leadership in their areas to constantly promote improvement to the quality of AM products and services.

When dealing with the customers, the top management needs to ensure that all the requirements specified by the customer are met consistently on top of the statutory and regulatory requirements. As mentioned in Section 1.1, despite AM not being well established at present, having control over raw materials, process guidelines, material testing, etc., improves the confidence level of customers. In addition, the AM organization must address the different risks involved in the fabrication of the product and ensure that conformity is met. It is also necessary for the top management to maintain a good relationship with the customer and improve customer satisfaction with the products and services they deliver.

In order to achieve the aforementioned, the top management is required to establish, implement, and maintain a set of quality policies suitable to the AM context and direction of the organization and define a framework based on AM needs to implement objectives for quality. The AM organization's framework should include commitments to both the applicable requirements and continual improvement to QMS. The quality policies developed have to be properly documented and communicated to all stakeholders within the organization. Policies can include best practices while dealing with powder bed equipment, extrusion equipment, AM material storage requirements, etc.

As managing a large company can be challenging, the top management has to delegate the roles and responsibilities within QMS to relevant personnel and ensure that their roles are well understood and communicated in the organization. The quality policy of an AM organization could adopt the following [6]:

- Ensure that the QMS conforms to ISO 9001:2015.
- Ensure that AM processes are delivering their intended outputs.
- Report the performance of the QMS and seek chances to continually improve the system.
- Ensure that the QMS is customer orientated.
- Ensure that the integrity of QMS is well maintained if there are any changes made to the system.

1.3 Planning

The organization needs to understand the requirements and expectations of interested parties in AM fabricated parts. Prior knowledge of the needs allows the organization to identify the risks and opportunities that have to be addressed, as part of implementing a QMS to meet customer needs. The organization has to ensure that their roles are well understood and communicated within themselves, as well as to the customer. For example, the customer would have to be aware of the inherent risks associated with AM, arising from factors, such as material or process limitations. The organization has to assure all stakeholders that the QMS can attain desired results, improve intended effects and decrease unwanted effects, and strive to continually improve on the system. As the QMS may still be undergoing a developmental state, continual improvement can be achieved through proper actions planning to address risk and opportunities, integration of QMS, and evaluation of the usefulness of the actions taken. The actions have to be proportional to the impact on the organization's products and services.

The AM organization should adopt ISO 9001:2015 guidelines to delegate the planning of quality objectives to people at different levels, functions, and processes that they are involved in. The AM quality objectives have to be in line with the policy, measurable, able to meet requirements, and conform to printed products and services desirable for the customers. If a customer wants to print 10 sets of a product, the organization must monitor the output results and ensure that all the 10 printed parts meet the specifications given by the customer. The products should conform to the requirements, and if there are any errors, the organization has to find out the root cause of the problem. The objective results have to be monitored, communicated to relevant stakeholders, and updated when necessary. All this information on quality objectives has to be documented and maintained.

If there are any changes required to the QMS, the considerations to be included are the purpose of change and their consequences, resources, and manpower allocation and other factors that may affect the AM organization's stand in quality.

1.4 Additive Manufacturing Support

To ensure proper implementation, maintenance, and improvement of the QMS, the organization has to provide the resources needed to relevant stakeholders. They have to consider the constraints on internal resources, and if necessary, procure resources from external vendors. Examples of the resources are people, infrastructure, environment for process operation, resource monitoring equipment, traceability, and knowledge. In Singapore, the government is willing to cofund organizations with the resources to equip them with AM capabilities, through the National Additive Manufacturing Innovation Cluster (NAMIC) program. The program enables organizations to tap on grants from the government to train and equip their staff with AM capabilities and to carry out research and developmental work related to AM.

Human resources have to be managed through proper delegation of staff trained in AM handling. These trained staff will ensure that processes are implemented correctly. Additionally, the organization has to determine the necessary infrastructure required, such as buildings and utilities, equipment, software, and logistics resources that are needed to ensure that the products and services attain a certain standard. A comfortable and safe environment is also required for the staff to carry out activities, such as powder preparation, fabrication, postprocessing through reduction of noise level, good ventilation, etc. In the AM shop floor, high technology equipment is used and the safety of the staff is very important. Any staff member, regardless of roles, such as AM machine operators, engineers, or even cleaners, will be exposed to a degree of danger as long they are in the vicinity. Some of the hazards identified are:

- Corrosive chemicals used in postprocessing.
- Powder cloud due to AM material preparation.
- Exposure to high power laser.
- Exposure to extremely heated chamber.

Resources are required to monitor and verify that the system in place ensures that products and services meet a certain requirement. The organization has to ensure that the resources provided for monitoring are suitable for the process. If defective monitoring devices are used, it may result in an inaccurate result. All monitoring systems implemented by the organization have to be traceable to ensure confidence in the products and services. The monitoring equipment requires calibration from time to time and is traceable to national or international standards, and when such standards are not available, traceable to calibration documents. Additionally, for the safe operation of machinery, prevention of misuse or mishandling of material, and proper security measures, such as closed circuit television monitoring are needed for the area.

Resources are also required for knowledge retention of necessary processes and operations to ensure products and services are in compliance. The knowledge must be well maintained and made available easily too relevant personnel, to prevent loss of knowledge when people leave the organization.

The competence of relevant personnel is important to ensure the conformity of products and services. This will require resources from the organization to determine that a person operating the AM process is competent to ensure the effectiveness of the system. Some basic competencies consist of education and training for operating the process or service, evaluation of the personal implementation, and effectiveness of skill sets, which has to be documented as evidence for traceability.

The organization has to list the requirements of the QMS objectives, benefits, and effectiveness to all staff. On top of that, they have to warn the staff of the consequences of not conforming to QMS.

Documentation is important when it comes to the implementation of QMS. Regular updating of documents is required for the latest and most relevant results from operations and processes. Such documented information is also required for certification according to of ISO standards.

1.5 Additive Manufacturing Operation

The operation of any AM process has to be planned by the organization to ensure that the products and services are in accordance with the defined requirements. The organization needs to set up the criteria for all AM processes, establish acceptance level of products and services, determine the resources required for the operation, ensure proper implementation of process control, and document all relevant information necessary for the process to function with conformity. Some examples of criteria for processes are printing parameters for a certain metal powder and postprocessing duration for removal of support material. The acceptance level of the printed part has to meet requirements, such as dimension tolerances and surface finish that is required by the customer. Moreover, the AM organization can prove that their process is optimized and stable and the product quality requirements are consistently met. As part of the operation, the AM organization can refer to the current standards published by organizations, such as ISO or ASTM International that relate to materials, processes, and test methods for qualifying the fabricated products.

Additionally, to ensure that products and services meet the defined requirements, the organization is required to establish proper communication channels with customers. Information regarding products and services has to be made known to the customer, while enquiries, contracts, orders, and customer feedback related to products and services are to be well handled. Transparency between both parties is also necessary for establishing trust in the recommended AM process. The organization has to as certain proper handling of customer property and if necessary, plan for contingency action. The product and services offered to customers have to meet both statutory and regulatory requirements, if applicable.

As part of the adaptation of ISO 9001, the AM organization has to conduct reviews of their processes, products, and services to establish that they meet the following requirements [6]:

- Requirements by customer.
- Requirements by organization.
- Requirements by statutory and regulatory bodies (if applicable).
- Contract/orders requirements.
- Requirements not specified but necessary for certain situation when the information is made known.

All these requirements have to be made known to and agreed by the customers before acceptance of contract. This information has to be documented and retained by the organization.

If there are any changes made concerning the requirements for products and services, then all relevant documents must be amended and relevant personnel must be informed of the change in requirements.

The AM organization has to develop, establish, and implement a design and development process for their products and services. To determine the requirements for design and development, the organization needs to consider the following items which are similar to ISO 9001 [6]:

- Nature, duration, and complexity of design.
- Process stages to achieve the final product.
- Product verification and validation.
- Statutory and regulatory bodies that will be involved with the design process (if applicable).
- Resources, both internal and external, for the design process.
- Control interfaces required for the design process.
- Involvement of customer and user for the design process.
- Requirements for provision of products and services.
- The level of control in the design work as required by the customer and relevant parties.
- Proper documentation to show that the design requirements are met.
- Consequences that will be incurred if the product fails.

The AM organization shall implement a design and development control system to ensure that all the requirements are met. The following items are to be considered when developing a control system [6]:

- Description of the results that are considered acceptable.
- Conducting timely reviews to evaluate the design process to meet requirements.
- Verification and validation to ensure that the design process results in a product that meets the requirements for the intended use.
- Actions that are necessary after reviews, verification, and/or validation activities.
- Proper documentation and retention of information obtain during this control process.

All outputs from the design and development process have to meet requirements set by the organization and the customer, and also be adequate for subsequent process for provision of products and services. To meet the acceptance criteria, proper monitoring, and measurement requirements have to be set. The products and services have to be specified to meet their intended purposes and also be safe to use. All outputs from the design process have to be documented.

The AM organization may not have the capabilities to conduct all processes and resort to outsourcing some work to external vendors. If the organization decided to do so, then their vendors have to also meet the requirements set by the customer. Appropriate controls have to be determined by the organization when the products and services from external vendors are intended for [6]:

- Incorporation into the organization's products and services.
- Products and services for customers.
- Processes or part of a process that has been outsourced by the organization.

Therefore, all relevant criteria have to be controlled by the organization. Any processes performed by external vendors have to be considered in the QMS, and the vendors need to consistently meet the requirements set by customers and statutory or regulatory bodies, if applicable. Outsourced processes in the context of AM could include removal of printed parts from the base substrate, support material removal, heat treatment, surface treatment and finishing, and so on. The organization needs to review the effectiveness of the controls that are applied to the external vendors from time to time. Furthermore, all this information has to be communicated to the external vendor before they commit to provide any products or services to the organization.

To achieve certification for QMS, the products and services offered by the organization will have to meet the following requirements [6]:

- Documented information that defines the characteristics of products, services, and activities that are to be implemented, and the targeted results.
- Availability and use of proper monitoring and measuring equipment.
- Implementation of monitoring and measuring activities at stages to check that all the criteria for processes, outputs, products, and services are met.
- Proper use of resources like infrastructure, and so on.
- Qualified personnel to operate on the processes.
- Validation of processes.
- Plans to reduce or prevent human errors.
- Product release, delivery, and postdelivery implementations.

Identification and traceability mechanisms are necessary to ensure products and services conformity. There is a need to uniquely identify the statuses of all outputs throughout the production process. All documents necessary for traceability must be retained by the organization.

The organization must identify, verify, protect, and safeguard the property of the customers or external vendors when working on their property or their products. If a product is missing or damaged, the organization has to report it to the customer or external vendor and retain any necessary documentation related to the incident.

During product delivery, the organization has to make sure that the product is easy to identify and proper handling and control measures to ensure the proper delivery of the product are in place. They usually consist of [6]:

- identification,
- handling,
- contamination control,
- packaging,
- storage,
- transportation, and
- protection.

After the delivery of the product to the customer, the organization has to determine and meet the postdelivery requirements, for example, warranty, maintenance services, recycling, disposal of product, etc., that cover the following [6]:

- Statutory and regulatory compliance (if applicable).
- Potential unwanted results from products and services.
- Nature, use, and lifespan of products.
- Customer requirements.
- Customer feedback.

For any changes that are made to the operation of fabrication of products for the customer, the organization has to ensure that the changes conform to the requirements, and documents are updated, reviewed, and authorized by relevant appointed personnel.

To release products and services, the organization can only proceed if all arrangements have been satisfied and approved by the relevant authority if applicable, and if necessary, approved by the customer. All documentation related to the release has to be documented and must contain evidence of conformity and traceability.

If any nonconforming products are found, the organization needs to have controls in place to ensure that they are not delivered to or used by the customers. In such cases, the organization has to deal with non-conforming products through [6]:

- Informing the customer of the nonconforming product.
- Stopping current production and request return for delivered products.

- Correcting the process.
- Obtaining acceptance authorization under concessions from relevant parties.

All information related to the nonconformance, such as the description of the issue of non-conformity, actions required, concessions that are obtained from relevant parties have to be documented, and the relevant authority in the field of the nonconformity should be identified and informed. ISO 9001 serves as a fundamental basis for the adoption of a QMS. All procedures identified in this section are essential for the development of the system.

1.6 Additive Manufacturing Performance Evaluation

The performance and effectiveness of the QMS has to be evaluated by the organization, and all information arising from the evaluation of the QMS is to be documented as evidence for audit purposes. For example, an AM organization has to determine the parameters to be monitored and measured, methods of monitoring, measurement, results analysis, timeframe, frequency of monitoring and measurement, and also the time needed for result evaluation. This is because each organization has its own unique set of parameters, such as build temperature, laser power, laser speed, and powder size, therefore, for example, the building time and geometrical accuracy of printed products can be used as parameters for performance evaluation.

To measure customer satisfaction, the AM organization has to understand and objectively monitor what is perceived satisfactory by the customer. They also have to determine the techniques to obtain information from the customer.

With the collected data obtained from performance evaluation, the AM organization should evaluate and analyze the results to assess the following which are adopted from ISO 9001 [6]:

- Conformity of printed products.
- Customer satisfaction.
- Performance and effectiveness of QMS.
- Effectiveness of planning AM process.
- Effectiveness of actions taken to address risk and opportunities.
- Performance of external vendors.
- Improvement for QMS.

To ensure conformity, the organization requires routine internal audits at planned intervals to gauge if the QMS is effective and conforms to the requirements set by the organization and/or by the ISO standard, if applicable, and also if the QMS is implemented and maintained. For example, if an AM organization plans an internal audit program, they should adopt the following considerations:

- Frequency of audit.
- Methods of audit.
- Responsibilities of personnel involved in audit.
- Planning requirements and reporting.
- Audit criteria and scope.
- Selection of auditors and ensuring impartiality of audit process.
- Results of audit reported to relevant management.
- Taking corrective actions as soon as possible.
- Retaining information from the audit as an evidence of audit implementation.

The top management is also required to review the QMS to ensure that it is aligned with the direction of the organization. This review has to be conducted at planned intervals.

1.7 Improvement

Continual improvement is necessary for the QMS to enhance and improve customer satisfaction. The organization can achieve better customer satisfaction through the improvement of products and services, correct, prevent, and reduce processes that result in bad effects, and improve the QMS.

If there are any nonconformities arising from any AM fabricated product, the organization will have to act and rectify them, and bear all consequences arising from the nonconformity.

To deal with nonconformity, the organization shall [6]:

- Review and analyze the issue.
- Determine the causes of the nonconformity.
- Determine if similar nonconformities exist.
- Implement actions if necessary.
- Review the usefulness of the rectification.
- Update risk in planning if necessary.
- Amend and update changes in the QMS if necessary.

Information regarding the QMS should be documented and retained as evidence of the rectification actions taken.

1.8 PDCA Framework Cycle

There are many different types of framework that have been employed throughout the world. For an introduction, a simple framework that can

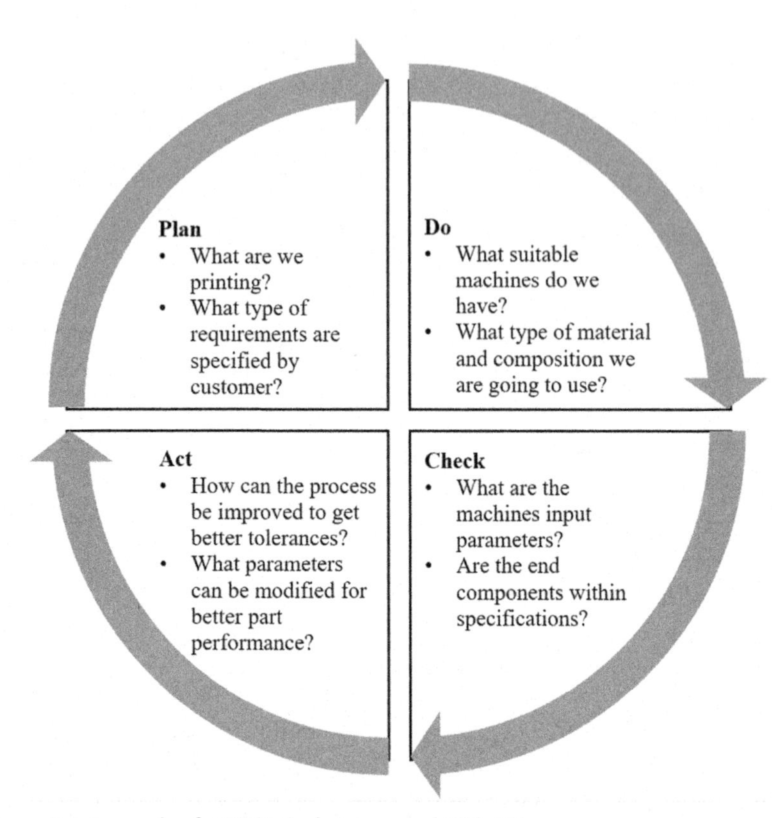

Figure 9.1 ***Example of a PDCA Cycle Integrated With AM.***

assist an organization in planning their QMS is a plan-do-check-act cycle (PDCA). A framework developed for this purpose will be further discussed in detail. The PDCA consists of four areas, and in the context of this book it has been adapted to suit AM (Fig. 9.1).

The PDCA cycle is as follows:

- Plan by establishing the objectives of the system and processes in AM, deliver results that meet customer requirements and identify and address the risks and opportunities.
- Perform printing and ensure all printing parameters are set for the correct material. Material handling is also important during the process to ensure no contamination.
- Check and monitor processes, products, and services against relevant company policies, objectives, requirements and standards, and report results.
- Act on what is reported and improve performance if needed.

The implementation of the PDCA cycle improves the company portfolio and customers' confidence. It results in faster turnaround time for acquiring raw material, producing higher quality products, and ensuring faster delivery to customers. Defects during product fabrication can be reduced thus resulting in lesser reworks, in turn saving both time and cost that will be incurred if a new part has to be fabricated again. PDCA implementation will also improve organization products and processes through continual improvement, potentially giving the organization a competitive edge against other suppliers.

2 ROLES OF REGULATORY AND CERTIFICATION BODIES

Regulatory agencies are governmental bodies created by a legislature to implement and enforce specific laws in their field, putting the public in their interest. The laws created by the agencies only apply in their country. Any products and services for the public and private sectors will have to be checked and tested before approval. The primary reason behind all the testing is to ensure that all the parts built are safe for use or consumption, and will not adversely affect the environment and health of people in any way if it was to be used. In the electronics industry, all mobile phones and radiotelephony equipment would have to be approved by The Federal Communications Commission (FCC) before it can be sold in the United States of America (USA), and if that particular mobile phone is to be sold in Singapore, though it may be certified by the FCC, will have to be recertified in Singapore by its local equivalent, Info-communications Media Development Authority (IMDA) [7]. This is to ensure that the radio frequency emitted by these devices is within a certain limit and criteria defined in local statutes and regulations.

Most additively manufactured parts at present are usually not certified by a regulatory body. However, some industries require stringent testing and part qualification due to the nature of their work. Aerospace, automotive, dental, and medical have very high standards of testing and part qualification, and therefore regulatory agencies have to set the standards for companies to meet the minimum safety requirements.

In the USA, governing agencies, such as the Federal Aviation Administration (FAA) and the Food and Drug Administration (FDA) deal with the qualification of parts made for aerospace, and medical, dental, and food, respectively [8]. Only if the parts have been certified by the relevant agencies can they be sold in the market. In Europe, many products are expected to be certified by an accredited body to conform to relevant European standards that enable a Conformité Européenne (CE) marking before the product can be introduced to the market. Both the FAA and FDA are exploring and determining the potentials and risks of AM in their relevant industries [9].

The FDA is an organization that protects public health through control of drugs, biological products, medical devices, food supply, cosmetics, and devices that emit radiation in the USA [10]. Under the medical sector, FDA assists research and innovations to make medicines effective, safe, and affordable for the public. FDA also ensures the security of food supply in the USA and fosters the development of new drugs and medical devices as part of the role in fighting against terrorism. In the AM industry, FDA regulates 3D printed drugs and medical equipment intended to be deployed in the medical industry in the USA, ensuring that it is safe for public use.

AM is widely explored in the area of customizing implants for patients. AM reduces the need to procure specialized machines to fabricate implants, thus lowering cost and making implants more affordable for patients. However, there is extremely limited confidence of use due to lack of proper certification of those parts to ensure biocompatibility. Thus, the FDA provides mechanisms to certify these implants through proper testing, in which they lay down the requirements of such implants to meet regulations. They have identified three requirements specifically for the medical industry, which are [11]:

- biocompatibility,
- mechanical properties, and
- interactive design.

The additively manufactured implants must be biocompatible to reduce the chances of biological rejection or the possibility of causing harm to the patient. They are also required to withstand forces that a human exerts, and yet be reasonably light enough for the patient to move about. Furthermore, the part has to be designed in a way where it is customized for that particular problem it is intended to solve. Additively manufactured implants are designated as class III devices generally reserved for devices considered as the highest risk, such as replacement of heart valves, and typically require approval before they can be marketed. The FDA regulates the entire process from design to the implant stage to ensure that the device implanted will not harm the patient. It is also consulted on all aspects including design, raw material, printing process and procedures, cleaning, and sterilization before an approval is given to proceed with the works [11]. It is important that all medical devices produced by AM must be safe and effective for usage [12]. These devices must go through thorough checks and tests to ensure that they will not harm the user or patient.

In Aug 2015, FDA approved a new form of 3D printed drugs (pills) that will pave the way for future development in medical drugs. "Spritam," a pill sold by Aprecia Pharmaceuticals allows easier consumption for patients who have issues swallowing through a unique structure design which allows the drug to dissolve faster [13,14]. The dosage could be more precisely controlled due to AM which aids in customization for selective treatments [13,14].

To give a perspective on the outlook of AM in the medical industry, FDA has approved about 85 such medical devices made via AM. These devices are not new or special. Components, such as hearing aids and dental devices have been produced by AM, which was previously produced by conventional manufacturing. The Centre for Devices and Radioactive Health (CDRH) mentioned that FDA views AM similarly to Computer Numerical Control (CNC) machining, where both technologies are used to fabricate medical devices [15].

The FDA is also in charge of approving AM fabricated parts for use in dentistry. Dentca Inc., a company producing dentures, prints the dental bases via the SLA AM process. These dental bases have satisfactorily passed the required tests in accordance to FDA blue book memorandum #G95-1 and ISO 10993-1 [16]. This technology speeds up the production by 2.5 times as compared to conventional methods. Errors in the dental bases are also reduced as there are far less processes and fabrication time was reduced to 5 days from a previous 30.

The FDA plays a huge role in regulating medical devices in the USA as they will ensure medical devices made by AM are high quality, fast in production, safe, and cheap to use. Approvals by FDA will also boost both consumer and industry confidence in the AM sector, which in turn generates revenue for the AM industry. Since 2015, there are about 20,000 acetabular cups produced by EBM that have been approved by FDA and CE [17]. The growth in manufacturing of implants through AM has resulted in gaining larger interest from approving authorities, resulting in easier clearance, and approval.

In aerospace, the FAA is the national aviation authority of the USA, which regulates and oversees all aspects of civil aviation in the country [18]. The FAA is divided into four lines of business and they are:

- airports,
- air traffic organization,

- aviation safety, and
- commercial space transportation.

AM is mainly applicable to two lines of business: aviation safety and commercial space transportation. All commercial based aircraft or spacecraft with components fabricated by AM will have to be approved by FAA before the vehicle is allowed to fly. Parts have to be evaluated for airworthiness through proper fabrication and ensuring that all the process involved in making the part meet requirements. To address the needs and concerns, FAA formed a working group, Additive Manufacturing National Team (AMNT), to research and develops the potential of AM in the engine and airframe design, metallurgy, inspection, and general aviation.

Furthermore, AMNT engages other USA federal agencies and academia in collaboration for developing and establishing guidelines for certifying AM products [19]. This collaboration is expected to accelerate the integration of AM into the aerospace industry.

Despite the many benefits AM can provide the lack of understanding of many variables may lead to more risk [19]. Unlike traditional methods which have a well-documented history of the effects on the material, there is very little knowledge on AM alloys and their properties. This is due to nearly 120 variables that need to be controlled for a given machine to produce stable and repeatable parts [20]. There is always a need for testing to prove that the parts produced through AM are suitable and safe for usage throughout the lifespan of the part.

Under the FAA design approval list, AM is classified as a fabrication method, for both metal and non-metal, where raw materials are processed in a machine to produce near net shape or near final parts. AM parts have to be in compliance with sections 25.603, 25.605, and 25.613 to be approved by FAA for use in the aerospace industry. Also, all process parameters, raw materials used, and machine key characteristics have to be logged and documented. This is to ensure traceability and repeatability of parts produced by AM.

2.1 Section 25.603 Materials

This section describes the suitability and durability of materials used for parts, which in failure will adversely affect safety. The part must be established by experience or test, conform to approve specifications and taken into account on how environmental conditions will affect the part [20].

2.2 Section 25.605 Fabrication Methods

This section describes the requirement of fabrication methods to produce a consistently sound structure. To achieve a sound structure, the fabrication process that requires close control must be performed under an approved process specification, and each new aircraft fabrication method must have a test program to ensure airworthiness [20].

2.3 Section 25.613 Material Strength Properties and Material Design Values

This section describes the design requirement and material strength for aerospace parts. The strength properties of a part must be based on testing of materials to meet approved specifications. Design values must be chosen to minimize the probability of structural failure, where single load path structures must meet a 99% probability with 95% confidence statistics and redundant load path structures must meet a 90% probability with 95% confidence statistics [20].

Testing and proving of components for the aerospace industry require many layers of approval, which has resulted in components fabricated by AM facing difficulty in getting approvals for use. Nevertheless, milestones have been met with a newly fabricated component by AM which has been certified by FAA—a temperature sensor housing 3D printed by General Electric (GE). The use of AM had reduced their design time by a year, saving cost in the long run, and also allowing faster implementation of new technology in the industry. The housing will be used in the next generation of LEAP engines created by GE [21–23]. This marks one of the first 3D printed parts being certified by FAA and it paves way for future parts to be approved and used in aerospace applications.

Having government agencies that approve AM parts for use in the industry will expand the boundaries of AM. Adoption rate will increase as confidence in AM increases, which can only be achieved through proper certification and qualification of AM parts. Therefore, testing and proving the efficacy of AM parts is crucial [8].

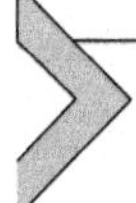

3 PROPOSED FRAMEWORK FOR ADDITIVE MANUFACTURING IMPLEMENTATION

AM standards that have been developed in recent years only address certain topics [24]. The framework will discuss on the process of implementing quality beginning from the initial design work, equipment

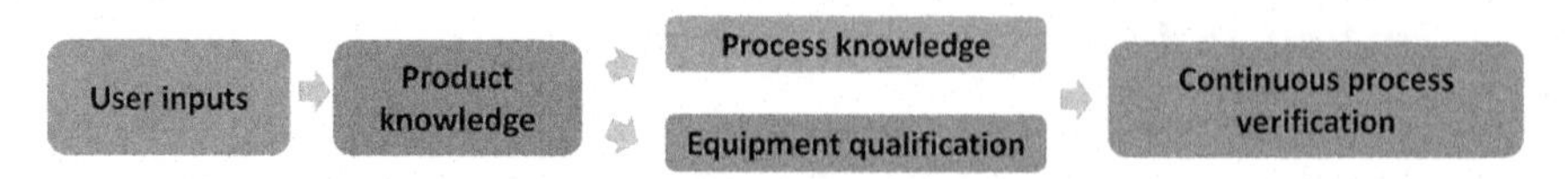

Figure 9.2 ***Proposed Framework for AM.***

processes that is needed for fabrication of the part, final part verification, and continual improvement on the quality system.

With the standards serving as foundations, an AM framework is proposed (Fig. 9.2) to cover five main areas:

- user inputs,
- product knowledge,
- equipment qualification,
- process knowledge, and
- continuous process verification.

3.1 User Inputs

3D data files from computer-aided design (CAD) software and scanned files from 3D scanners are the basic ingredients required for printing of parts with AM equipment. The current defacto standard file format widely used in the industry is STereoLithography (STL), which acts as a container to transfer data between CAD programs and AM equipment. STL natively has inherent problems—being a surface mesh-based file format, STL uses triangle elements to generate shapes of different dimensions. Although it does not usually have problems replicating flat planes and surfaces exactly to the same dimensions as compared to native CAD data, problems arise when there is a need to replicate curved surfaces. A large triangle element with low mesh density will result in tessellation, thus not representing a curved surface accurately. In order to achieve better accuracy, smaller triangle elements need to be used along with a high mesh density to more accurately represent curved surfaces at the expense of data space and computing power. The STL format also does not contain other relevant information, such as color, texture, material properties, and so on, to accurately represent the part.

The limitations of STL have led standardizing bodies and companies to develop new file formats specifically for AM. One new standard developed by ASTM is the Additive Manufacturing Format (AMF). The new AMF format will be able to support advanced features like multimaterial support, color information, functionally graded materials, and so on. Another file format initiated by the industry is the 3D Manufacturing Format (3MF), developed by the 3MF consortium. Driven primarily by requirements from the industry, big players, such as Microsoft, Stratasys, Autodesk, HP, 3D systems, Siemens, and others form a consortium with an aim to address key issues that arise from working with STL.

A more recent development in AM equipment software accepts most CAD file formats natively without the need of converting them into the STL format. For example, Stratasys recently collaborated with GrabCad, an online CAD file repository, to develop a 3D printing program that imports most native CAD file formats, translates it internally, and sends the part for printing directly to compatible Stratasys machines. Siemens NX has the capabilities of printing direct to DMG Mori Lasertec 65 3D, a hybrid AM equipment through an integrated CAM module. All these advancements in software could eventually render the de-facto STL format obsolete, but the transformation is still expected to take years or even decades.

Other sources of 3D data format files are from 3D scanned data, or multiple 2D scan data stacked together in succession to form a 3D model. There are multiple scanning technologies in the market, with the most popular ones using dual cameras to triangulate points on a surface of an object, before translating it into point cloud data. The user would just have to place the object onto the 3D scanner, and at a press of a button, a 3D model will be generated into the computer. However, issues like noise, reflective surface, accuracy of the scan, etc., frequently cause problems with the 3D model. One could also question the accuracy of the 3D model, and how well it compares with the actual object itself.

3.2 Product Knowledge

AM enables designers to design with greater freedom with regard to the consideration of the limitations of the manufacturing processes. Parts of different sizes and shapes can be fabricated within the same machine at the same time. Therefore, there is a need for a designer to understand the product well enough to foresee and address all the risks involved from the conceptual stage to the end product.

Design for AM differs from design for conventional manufacturing, as it allows the design for complex geometries, functionally gradient design in terms of shapes or materials, reducing assembly through the fabrication of integrated parts, and consolidation of parts into each production run [25].

There is a need to develop new design methodologies to optimize the product for an AM process. The designer has to understand the available processes and the inherent support generation that comes with that process. Processes like fused deposited modelling (FDM), selective laser melting (SLM), stereolithography apparatus (SLA), digital light projection (DLP), and material jetting will require support generation for the part with overhangs for print stability. These structures not only function as anchors but also help to dissipate heat and prevent thermal warping of the parts [26]. Unfortunately, these support structures may be difficult to remove after a print, especially if they are made of metal, which will require machining to remove the support. On the contrary, SLS does not require support structure. In SLS, the loose powders acted as supports which can be easily removed once the print is completed. Therefore, it is very important to know what process will be adopted in the production of the designed parts, and consideration for support structures must be taken for ease of postprocessing.

Addressing the risks associated with a product is also required right from the design stage. With a good understanding of the product requirements and targeting the correct risks through risk assessment, it is possible to reduce the number of validation experiments that are required to be conducted. In turn, it reduces overall time and cost incurred by the company.

3.3 Equipment Qualification

Any AM equipment will require qualification in three areas—the software used on the equipment for printing, the performance and characteristics of the equipment, and process materials, such as coolant and oil, which the equipment uses for fabrication. All these have to be traceable according to ISO 9001:2015 requirements. AM equipment would also require proper and timely maintenance and calibration following the AM manufacturer specifications, and calibration to a certain standard, if applicable.

Software that comes with the equipment, together with any other 3rd party software used for the processing and generation of the printing sequences, should be validated. The software has to be designed to demonstrate that the equipment and the program controls will perform all intended functions before the software is allowed to be deployed for part production.

The performance of any AM system can be assessed by fabricating a standardized test artifact. NIST has proposed a test artifact to investigate the performance and capabilities of an AM system that was discussed in Chapter 3. AM systems using the similar process, have to be able to produce the same parts continuously, regardless of the machine age. The research for

AM machines is on-going, and more work is needed for an in-depth understanding of the processes.

The manufacturer of the AM system has to prescribe procedures and guidelines that ensure that their clients' or users' machines are always in good condition. This may consist of inspection, maintenance, and calibration of the machine that has to be responsibly performed by the user or client, who may in turn have to routinely engage the manufacturer for some of the tasks. Both the manufacturer and the user or client also have to ensure that themaintenance and calibration of the machine are performed by qualified personnel and all adjustments and maintenance activities are well documented.

Indirect process materials used during fabrication have to be removed from the end product. As these materials are not intended to be present in the final product, it has to be demonstrated by the manufacturer to regulating bodies and approving agencies that the product is free from indirect process material, and the product safety will not be affected by the use of process material. One such indirect process material is the gas used during fabrication. The melting of metal in a SLM process requires argon or nitrogen gas to flood the chamber to act as an inert blanket. However, these gases are not intended to be in the final design part and have to be removed when the print is done. Thankfully, the removal of these gases is relatively simple through purging of the chamber once the print is completed. It is also important that these process materials do not induce undesired effects on the part itself, such as altering the microstructure or chemistry of the part that may affect the overall safety of the product.

3.4 Process Knowledge

It is essential to understand the print process in order to predictably determine the end properties of the part, as well as foresee potential defects that may be induced by the process. Accurate and comprehensively validated process models are a practical necessity in advancing adoption and deployment of AM processes in real-world manufacturing [27].

Critical sub-processes, such as laser and optical systems operation, temperature control, motion control, and material deposition control need to be identified. Additionally, indicative parameters are to be standardized and monitored for each of these subprocesses. These parameters may be scan speed, layer thickness, scanning strategy, and so on. An understanding of how these parameters translate to the output quality attributes of the part, such as dimensional accuracy, surface roughness, microstructure, and mechanical properties needs to be achieved. It would then be possible to identify the key parameters, their optimum values, and control limits to achieve the desired output characteristics based on this understanding [24].

Management of materials is of fundamental significance in an AM process, especially on an industrial scale. Traceability of material is a prerequisite to assure the quality and integrity of material in a manufacturing environment. Present-day examples of methods to enhance traceability of AM materials include sealed FDM and Polyjet material cartridges from Stratasys with embedded radio-frequency identification (RFID) tags that can be read by the AM equipment. However, traceability is comparatively limited on powder-based processes of both metals and polymers. The situation is further complicated by the recycling of unused powder and mixing with newer batches of powder involved in these processes. There are studies that show adverse effects of recycled powder on the structure and mechanical properties of the part [28], or degradation of polymer powder over time with repeated heating and cooling [29]. Apart from ways to enhance traceability, such effects on the material should also be well understood.

Well-founded process understanding is vital to enable a closed loop feedback system with in situ continuous process verification for the subsequent phase of the proposed framework. Not only does this include the technical aspects of the process, but also the management of subprocesses, process parameters, and input materials that are identified as being critical to ensure that the quality of the parts is consistently upheld.

3.5 Continuous Process Verification

The fundamental nature of the AM process warrants the need for continuous process verification. As compared to a part machined out of a block of bulk material with largely uniform microstructure and mechanical properties, any minor abnormality in the raw material or that occurs at any point during the AM process that goes unnoticed may compromise the quality of the part. For example, in a laser metal sintering AM process, a slight flicker of the laser beam due to an electrical surge or an abnormally sized powder particle may alter the microstructure of the part at just one critical layer. This deviation may result in a catastrophic failure of the produced part if it goes unnoticed. Unfortunately, it would not be possible to reliably examine the part for such abnormalities after production, especially if it consists of complex geometry designed to leverage the benefits of AM.

Characterization of raw materials and determining their acceptable specifications is necessary to enable continuous process verification. Specifications for the raw material may consist of attributes, such as particle size, size distribution, morphology, viscosity, melting point, and nonintrinsic properties, such as powder flowability [24]. Monitoring of the material that is fed into the process is key to ensuring a consistent supply of material to the AM process.

In situ monitoring of the core process of AM, such as sintering or melting would be required to ensure part consistency, especially with the relatively long production times usually involved. This may be performed by means of an automated visual inspection system and also identifying and monitoring indirect environmental factors in the build chamber, such as excessive smoke due to incorrect melting.

The geometry of the part at a cross-sectional level could also be compared against the data from the digital model to ensure dimensional accuracy during the printing process. This would also enable geometric validation of internal features of a complex part that may not be possible to the same level at a later stage with traditional metrology instruments. Optical measurement of the parts at their cross-sections also aids in effectively assessing tolerances of any free-form surfaces in the product [30]. The process can be appropriately controlled by means of a feedback mechanism from the monitoring data based on the validated process models.

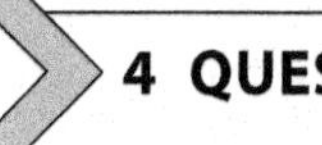

4 QUESTIONS

1. Explain the role of a QMS in an organization.
2. Why is continual improvement in a QMS important?
3. Distinguish between the roles of a "standards defining body" and a "regulatory or statutory body."
4. Discuss the importance for a designer to understand the different AM processes.
5. Discuss the importance of well-founded process understanding and its relevance in enabling continuous process verification.
6. What is the current de facto file format used in AM? List some of its limitations.
7. Discuss the importance of equipment qualification from the perspective of AM.

REFERENCES

[1] C.K. Chua, K.F. Leong, 3D Printing and Additive Manufacturing: Principles and Applications, fifth ed., World Scientific Publishing Company, Singapore, (2017).
[2] C.K. Chua, M.V. Matham, Y.J. Kim, Lasers in 3D Printing and Manufacturing, World Scientific Publishing Company, Singapore, (2017).
[3] M. Mani, B. Lane, A. Donmez, S. Feng, S. Moylan, R. Fesperman, Measurement Science Needs for Real-Time Control of Additive Manufacturing Powder Bed Fusion Processes, National Institute of Standards and Technology, Gaithersburg, MD, USA, (2015).
[4] M.J. Harry, R.R. Schroeder, Six sigma: The Breakthrough Management Strategy Revolutionizing the World's Top Corporations, Broadway Business, (2005).
[5] R. Shah, P.T. Ward, Lean manufacturing: context, practice bundles, and performance, J. Oper. manag. 21 (2003) 129–149.
[6] ISO, ISO 9001:2015, quality management systems, ISO, 2015.
[7] IDA, Equipment registration framework. Available from: https://www.ida.gov.sg/EquipmentRegistrationFramework, 2015.
[8] 3D Engineer, Federal regulations for 3D printing. Available from: http://www.3dengr.com/federal-regulations-for-3d-printing.html, 2015.
[9] R. Wright, Regulatory concerns hold back 3D printing on safety. Available from: https://www.ft.com/content/bfab071c-6abc-11e4-a038-00144feabdc0#axzz3udU0khrB.html, 2014.
[10] FDA, FDA—What We Do. Available from: http://www.fda.gov/AboutFDA/WhatWeDo/default.htm, 2015.
[11] S. Leonard, FDA grapples with future regulation of 3-D printed medical devices. Available from: http://www.mddionline.com/article/fda-grapples-future-regulation-3-d-printed-medical-devices-140613, 2014.
[12] R.J. Morrison, et al. Regulatory considerations in the design and manufacturing of implan table 3 D-printed medical devices, Clin. Transl. Sci. 8 (2015) 594–600.
[13] R.J. Szczerba, FDA approves first 3-D printed drug. Available from: http://www.forbes.com/sites/robertszczerba/2015/08/04/fda-approves-first-3-d-printed-drug/, 2015.
[14] D. Basulto, Why it matters that the FDA just approved the first 3D-printed drug. Available from: https://www.washingtonpost.com/news/innovations/wp/2015/08/11/why-it-matters-that-the-fda-just-approved-the-first-3d-printed-drug/?utm_term=.c672880d4781, 2015.
[15] J. Hartford, FDA's View on 3-D printing medical devices. Available form: http://www.mddionline.com/article/fdas-view-3-d-printing-medical-devices, 2015.
[16] E. Krassenstein, DENTCA receives FDA approval for world's first material for 3D printed denture bases. Available from: http://3dprint.com/87913/dentca-fda-3d-print/, 2015.
[17] D. H. Trinh, Regulatory approval of implants produced with additive manufacturing, presented at the 13th annual Orthopaedic Manufacturing & Technology Exposition and Conference (OMTEC 2016), Chicago, IL, USA, 2015.
[18] FAA, Federal aviation administration. Available from: https://www.faa.gov/, 2015.
[19] T. Hoffmann, Your airplane is ready to print! Available from: http://www.faa.gov/news/safety_briefing/2015/media/MayJun2015.pdf, 2015.
[20] J. Kabbara, FAA: Additive manufacturing, presented at the Gorham PMA and DER conference, San Diego, CA, USA, 2015.
[21] T. Kellner, The FAA cleared the first 3D printed part to fly commercial jet engine from GE. Available from: http://www.gereports.com/post/116402870270/the-faa-cleared-the-first-3d-printed-part-to-fly/, 2015.
[22] Metal Additive Manufacturing, GE aviation to retrofit over 400 commercial jets engines with new additive manufactured sensor. Available from: http://www.metal-am.com/ge-aviation-to-retrofit-over-400-commercial-jets-engines-with-new-additive-manufactured-sensor/, 2015.

[23] GE Aviation, First additive manufactured part takes off on a GE90 Engine. Evandale, OH, USA, 2015.
[24] W.Y. Yeong, C.K. Chua, A quality management framework for implementing additive manufacturing of medical devices, Virtual Phys. Prototyp. 8 (2013) 193–199.
[25] I. Gibson, D.W. Rosen, B. Stucker, Additive Manufacturing Technologies, Springer, New York, USA, (2010).
[26] M.X. Gan, C.H. Wong, Practical support structures for selective laser melting, J. Mater. Process. Technol. 238 (2016) 474–484.
[27] J. Pellegrino, T. Makila, S. McQueen, E. Taylor, Measurement science roadmap for polymer-based additive manufacturing, Material Measurement Laboratory, National Institute of Standards and Technology, 2016.
[28] M. Toth-Tas¸cău, A. Răduţă, D.I. Stoia, C. Locovei, Influence of the energy density on the porosity of Polyamide parts in SLS process, Diff. Defect Data Pt.B Solid State Phenom. 188 (2012) 400–405.
[29] K. Plummer, M. Vasquez, C. Majewski, N. Hopkinson, Study into the recyclability of a thermoplastic polyurethane powder for use in laser sintering, Proc. Inst. Mech. Eng. 226 (2012) 1127–1135.
[30] K. Wolf, D. Roller, D. Schäfer, Approach to computer-aided quality control based on 3D coordinate metrology, J. Mater. Process. Technol. 107 (2000) 96–110.

INDEX

D

E

K

L

M

N

O

P

V

W

X

Printed in the United States
By Bookmasters